CHILDREN: THE INVISIBLE SOLDIERS

CHILDREN THE INVISIBLE SOLDIERS

by Rachel Brett and Margaret McCallin

(Of the Quaker United Nations Office, Geneva, and
the International Catholic Child Bureau
on behalf of the Child Soldiers Research Project)

Rädda Barnen
SAVE THE CHILDREN SWEDEN

Rädda Barnen (Save the Children Sweden) is a non-governmental organisation that fights for the rights of children, in Sweden and around the world. Activities are designed to improve conditions for children at risk. Rädda Barnen acts by itself and in co-operation with others by:

• identifying and analysing problems and potential courses of action;

• sponsoring practical development and support programmes, and sharing the experience gained;

• influencing public opinion.

Rädda Barnen Publishing's books are primarily for people who work with children. Our goal is to disseminate knowledge concerning the situation of children and to provide guidance and impetus for new ideas.

ISBN 91-88726-53-3
© Rädda Barnen and the authors
Production management: Carl von Essen, AnnaLena Andrews
Editor: Anna Dahlin
Editorial assistant: Lisbeth Segerlund
Graphic design: Stefan Lundström, Neo Media
Cover: Stefan Lundström, Neo Media
Cover photo: Gamma/IBL
Second edition: 1
Printed by Grafiska Punkten, Växjö 1998

Contents

Preface

A s SHOWN IN THIS book, the child soldiers phenomenon appears in all corners of the world. South Africa has not been spared either. It has been awful to watch children dancing round a "necklaced" corpse.

It is immoral that adults should want children to fight their wars for them. Children should be playing, not being pawns in dangerous adult games.

From a Christian perspective, children bearing arms is particularly challenging, for the Biblical concept of the Kingdom is of universal peace, when they will beat their swords into ploughshares and their spears into pruning hooks and they shall learn war no more.

We must not close our eyes to the fact that child soldiers are both victims and perpetrators. They sometimes carry out the most barbaric acts of violence. But no matter what the child is guilty of, the main responsibility lies with us, the adults. Children are easily coerced into doing things they would never have done in a normal situation.

To stop this abuse, no child under 18 should be recruited for or participate in war – whatever the wishes of the child. There is simply no excuse, no acceptable argument for arming children.

The present book deepens our knowledge about an urgent issue, and points towards solutions. It is my profound hope that it will be read, in particular, by those whose decisions can make a difference for the individual child.

The Most Reverend
DESMOND M. TUTU
Archbishop Emeritus

Foreword

CHILDREN SUFFER GREATLY from participating in wars. They lose their childhood. They cannot play, they cannot go to school, and they are subjected to immense psychological and physical stress. Some are killed; many are maimed, tortured, or raped; and all of them are robbed of their fundamental right to a life in a safe and secure environment. Sexually transmitted diseases and drug problems are common. Child soldiers suffer serious psychological problems. Common symptoms are anxiety, apathy, nightmares, depression and concentration difficulties. Child soldiers are injured for life.

This book reveals that an increasing number of child soldiers are being recruited by governments and armed opposition groups, either by force, ideology, religion or promises of a secure and better life. The trend is becoming more evident. There are currently at least 300,000 child soldiers participating in armed conflicts.

Hundreds of thousands more are in peacetime armed forces which could find them selves in battle at a moment's notice. The majority of child soldiers are aged between 15 and 18, but a large number of children are recruited as early as 10. Each statistic represents a personal catastrophe, a human tragedy.

The book also shows that the increasingly common civil wars, fought in and among civilian populations, have dire consequences on children. When there is a lack of adult soldiers, children are recruited by government troops and armed guerillas, often brutally. However, children living in conflict regions often join in order to avoid violence to which they are regularly subjected by warring parties. A weapon provides access to food, and is better than staying home afraid and helpless. Children who join wars voluntarily are motivated by the social, economic and political conditions that rule their lives. Child soldiers are usually from the poorest and most vulnerable sectors. Armies and armed groups often view children as a cheap substitute when there is a lack of adult soldiers. Children are indoctrinated to be obedient; are easily manipulated and are considered cheap to keep.

Child soldiers are used for various kinds of service in addition to participating in military operations. These include guard duty, espionage, portering, cooking and providing sexual services. Technological advances have created the situation where a child with a lightweight automatic weapon can now cause as much injury as an adult.

The children never forget the cruelty they have experienced. Even those who are demobilised bear psychological scars which may have repercussions on society long after the demobilisation. Without rehabilitation and reintegration, child soldiers cannot readjust to society's norms and values.

Even if the picture painted in this book appears depressing, it is equally important to note the positive results of support work and international advocacy. For example, there are guerrilla groups and governments that have given in to international pressure and agreed not to use child soldiers. Family members and others have also openly and actively opposed children being recruited for wars. In addition, there are numerous examples of successful demobilisation, and rehabilitation projects in the world. One example is in southern Sudan where the SPLA guerilla has asked Rädda Barnen to help demobilise its child soldiers.

This book is intended for all those who are interested in the rights of the child in armed conflicts. The book is grounded in research by several organisations for the UN Study on the Impact of Armed Conflict on Children (the Machel Study). We would especially like to thank the organisations that carried out the 26 country case studies, and the authors, Rachel Brett and Margaret McCallin, for their excellent work of analysing the case studies and writing a highly readable and challenging book. Rädda Barnen has campaigned against the use of child soldiers for many years. We hope that this book will help us in our international work to stop the use of child soldiers. In closing, we would like to pay tribute to SAS, Scandinavian Airlines, for the support and funding that has enabled us to print the second edition of this book.

Suzanne Askelöf

Henrik Häggström

SECRETARY GENERAL

PROJECT MANAGER

Rädda Barnen

Rädda Barnen

Introduction

THE IDEA FOR THIS project was born in the NGO Sub-Group on Refugee Children and Children in Armed Conflict (one of the Sub-Groups of the NGO Group for the Convention on the Rights of the Child, in Geneva). A Steering Group was formed with the following members: the Quaker United Nations Office in Geneva, the International Catholic Child Bureau, Rädda Barnen (Swedish Save the Children), the Henry Dunant Institute, the Lutheran World Federation, World Vision International and the research component of the UN Study on the Impact of Armed Conflict on Children (the Machel Study). The Quaker UN Office led the project assisted by the International Catholic Child Bureau.

Funding was provided by Rädda Barnen, the Machel Study, the Swiss Federal Department of Foreign Affairs, the World Council of Churches and the Lutheran World Federation. Major contributions in kind were made by the Quaker UN Office and its parent bodies (Quaker Peace & Service and Friends World Committee for Consultation), the International Catholic Child Bureau, World Vision International and the National Inter-Religious Service Board for Conscientious Objectors (USA).

The authors wish to thank all those who carried out the research, and offered advice, encouragement and assistance in many ways, in what was truly a co-operative venture involving many different partners. Particular thanks for major contributions go to Derek Brett (for Annex 1), Rhona O'Shea (who worked as a volunteer on the Child Soldiers Research Project) and to Penny McMillin of the Quaker UN Office. In this context, tribute must also be paid to the decision of the Quaker Meeting in Geneva, endorsed by the 1979 Triennial Meeting of the Friends World Committee for Consultation, to take up the issue of child soldiers and to Dorothea Woods (of the Geneva Meeting) and to Martin Macpherson (of the Quaker

UN Office in Geneva) who played a remarkable part in putting and keeping the issue on the international agenda.

The research comprised case studies of particular situations where children are or have been active participants in armed conflicts or have been recruited into armed forces. The case studies were undertaken by local non-governmental organisations (NGOs), international NGOs working in the field, government or other agencies, on the basis of questionnaires prepared by the Steering Group. Due to the attitude of the authorities or to a shortage of time and funding, it was not possible to cover all those situations in which child soldiers are known to have been involved: secondary sources have been used to fill some of the gaps.

Case studies were received for Afghanistan, Bhutan, Burma/Myanmar, Burundi, Cambodia, Colombia, El Salvador, Ethiopia, Guatemala, Honduras, Lebanon, Liberia, Mozambique, Nicaragua, the Occupied Territories (Intifada), Paraguay, Peru, the Philippines, the Russian Federation (Chechnya)[1], Rwanda, South Africa, Sri Lanka, Turkey, Uganda, the United Kingdom (Northern Ireland) and the former Yugoslavia. In addition, a number of governments provided information about their recruitment and deployment of under-18s.

Eleven of the case studies were in-depth ones, responding to both questionnaires A and B; the others either addressed questionnaire A only or were negotiated individually to address the particular situation in the country. The questionnaires are reproduced in Annex 2. The case studies themselves are available from the Rädda Barnen Project on Child Soldiers (see address in Annex 5).

The purpose of this book is to develop a better understanding of the causes and consequences of children's participation in armed conflicts, not to expose or stigmatise the policies or practices of particular governments or armed opposition groups. Therefore, the material is treated in a way

---

[1]   The question of child soldiers was included in H. Balian, "Armed Conflict in Chechnya, its Impact on Children" (Covcas Center for Law and Conflict Resolution, Arlington, VA, November 1995), report prepared for the UN Study on the Impact of Armed Conflict on Children.

which is intended to avoid, as far as possible, questions about the accuracy of the presentation of particular situations, incidents or details. Instead, the focus is on what emerges as a remarkably similar pattern across different countries, regions and cultures.

To some extent, this generalisation makes a virtue of necessity. The contract to undertake this project was signed in September 1995, with a deadline of April 1996 for submission of the report to the UN Study. During this period, the questionnaires had to be developed, the countries and case writers identified, the cases researched and written up, and the report written. All participants were aware of the limitations imposed by this timetable, especially the case writers who demonstrated great understanding and flexibility in adapting their original plans to the exigencies of the time available. It is clear, however, that no individual case study is a comprehensive analysis of the situation of child soldiers in a particular country or conflict nor, taken together, do they constitute a comprehensive coverage of the situation of child soldiers in the world.

The considerations outlined above have influenced the way in which the case study material has been handled in the text. Where the text of the case studies is quoted directly, it is duly acknowledged but, in general, wording which identifies the specific situation or names of individuals, groups or places has been edited out. No attempt at concealment is intended by this practice; the missing names will often be evident to those familiar with the country concerned. The aim is to avoid the general picture being obscured by the specific instances used to illustrate it.

Another feature of the style of treating the material arises from the same considerations. No distinction is made between time periods nor between actors unless specific contrasts are being made. None of the case studies deals with remote history; all of the experiences reported are from the recent past. In some instances practices are known, or believed, to have changed; a number of the conflicts have allegedly or actually ended; several have entered a new phase between the commissioning of the case studies and the completion of this book, but the nature of the experiences reported remains contemporary, therefore they are reported in the present tense.

DEFINITIONS

At certain points there are clear distinctions to be drawn between what for the purposes of this book are defined as "government armed forces" and "armed opposition groups". The former applies to all regular governmental, irregular governmental, government-supported or condoned forces, armed police or internal security forces, militias and self-defence groups[2] although, where possible and relevant, distinctions are made between them.

Similarly the term "armed opposition group" is used to refer to all other armed groups, even where more than one armed group was operating concurrently or consecutively, and even if the group in question subsequently became the government. The distinction between government and opposition relates to legal authority and the question of territorial control at the time in question. The authors of this book take no position about the merits of the conflicts or their causes, being concerned only with the nature and effects of the involvement of children, by any party and for any reason.

Graça Machel, the expert appointed to undertake the UN Study on the Impact of Armed Conflict on Children, has stated that 18 years should be the minimum age for recruitment and participation in hostilities, in line with the general age of majority contained in Article 1 of the 1989 UN Convention on the Rights of the Child. For the purpose of this project, this was, therefore, taken as axiomatic. Thus the term "child" is used for any person under the age of 18 years although, under Article 38 of the Convention on the Rights of the Child, the minimum age for recruitment into armed forces, as well as for participation in hostilities, is 15 years. The same age is stated in the 1977 Additional Protocols to the 1949 Geneva Conventions governing the conduct of armed conflicts. Under national legislation, the minimum ages for compulsory and/or voluntary recruitment vary: the phrase "under age" is used in relation to this

---

[2]   Except where there is clear evidence that these are neither government supported nor condoned.

14

recruitment age, or the age formally stated in the policy of the armed opposition group concerned. It should also be noted that account is taken of the situation of those now over 18 years but who fell within this age group at the time of their recruitment.

The term "soldier" is used for a member of any kind of regular or irregular armed force or armed group in any capacity and those accompanying such groups, even if not defined or identified as members, other than purely as family members; it therefore includes cooks, porters, messengers, and so on. However, a distinction is made between those who carry arms and participate in combat directly ("combatants"), and those serving in support capacities.

The Case Studies

Country	Years covered	Forces and Groups Studied
Afghanistan	1978–	(1979–92) Government forces; (1979–92) Mujahideen; (1993 –) various factions
Bhutan	1990-	Government Army (RBA), Police (RBP), Militia
Burma/Myanmar	*1988–	Government forces; Opposition groups: KNU/KNLA; ABSDF; NMSP/MNLA; USWP/USWA; PDF
Burundi	1993–	Government forces; Hutu and Tutsi Opposition groups
Cambodia	1970–	Successive Government forces; Opposition Khmer Rouge
Colombia	1965–	Government forces (army, paramilitary); Opposition groups (FARC, ELN, EPL, Quintin Lame, M19); "Popular Militias"
El Salvador	1980–1992	Government forces (FAES); Opposition (FMLN)
Ethiopia	1974–1992	(to 1991) Government forces (Derg); Ethiopian People's Revolutionary Democratic Front (EPRDF); (1992) Oromo Liberation Front (OLF)
Guatemala	1960–	Government forces (Army, PAC Militias); Guatemalan National Revolutionary Unity (URNG)
Honduras	*1977–	Government forces; unnamed Opposition groups
Lebanon	1975–	Government forces; sectional militias; South Lebanon Army; PLO; Hezbollah; Black September Brigades *etc.*
Liberia	1987–1993	Government forces; Opposition NPFL, ULIMO and other factions
Mozambique	1976–1992	Government (Frelimo) Army and allied self-defence teams/local armed groups, e.g. NAPARAMAS; Opposition RENAMO
Nicaragua	1977–1979	Government forces (National Guard)
Occupied Territories (Intifada)	1987–1993	Unstructured opposition
Paraguay	*1989–	Government forces (Army, Police)

Peru	1980–	Government forces (Army, Self-Defence Committees); Opposition (Shining Path, MRTA)
Philippines	*1976–	(Opposition) New Peoples Army; also (briefly) Moro National Liberation Front (MNLF)
Russian Federation (Chechnya)	1994–	Opposition groups
Rwanda	1994–	Government forces; Interhamwe Militias
South Africa	1961–	(to 1991) Government forces; armed wings of the African National Congress and Pan African Congress (post 1991) "Self-Defence Units" & "Self-Protection Units"
Sri Lanka	1983–	Liberation Tigers of Tamil Eelam (LTTE)
Turkey	1984–	AGRK – military wing of the (Opposition) PKK
Uganda	1989–	Lords Resistance Army (LRA)
UK (Northern Ireland)	1969–	Republican and Loyalist Opposition groups; unstructured opposition
the former Yugoslavia	1991–	"Army of Republika Srpska"; "Army of Republika Srpska Krajina"; informal Serbian groups

* In these cases, the date given is of the earliest first-hand testimony quoted in the case study, the recruitment of children having gone on for an unspecified length of time previously.

Notes:

The case studies are available from Rädda Barnen's Project on Child Soldiers (see address in Annex 5).

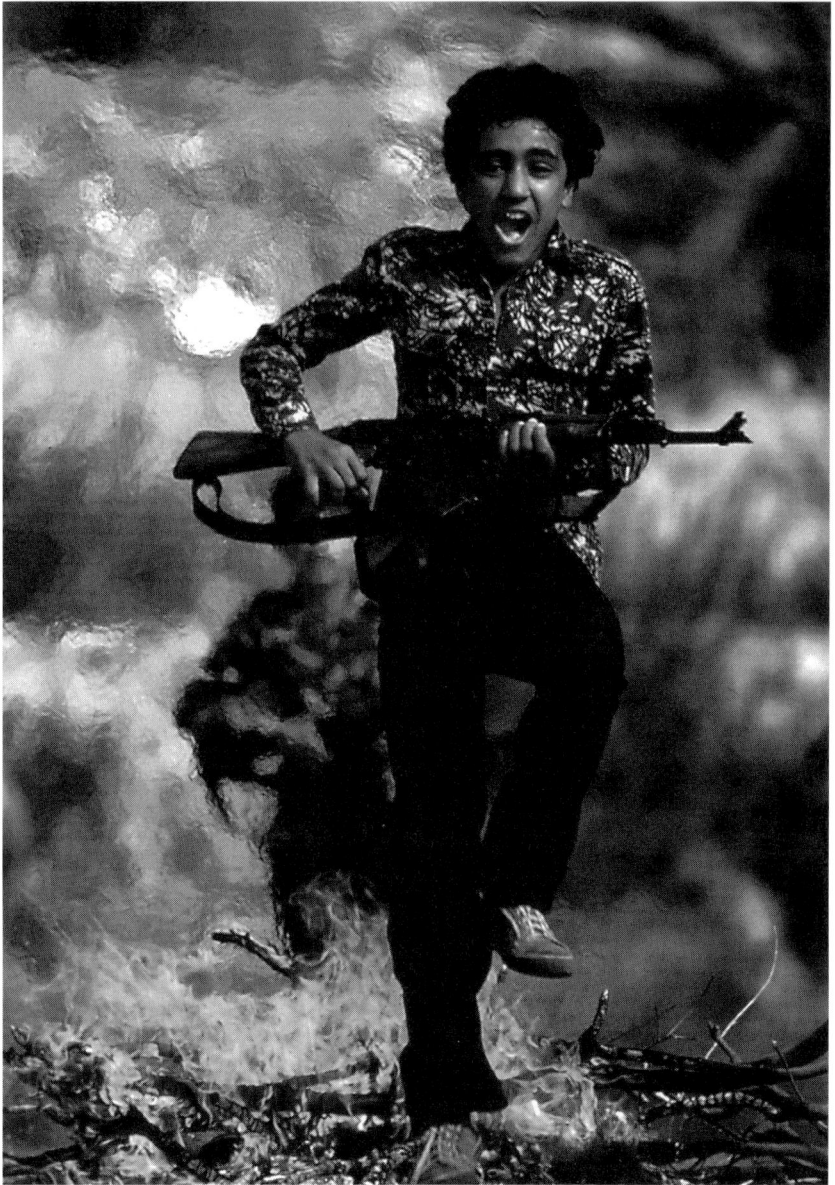

A young Palestinian in training. The development of light arms has made it easier for children to serve as combatants.
Photo: Carol Spencer, Gamma/IBL.

CHAPTER I
The Global Picture

A sizeable proportion [of the opposition forces] were below 18 – a fact that was much noticed when the [opposition] captured [the capital] ... The subsequent demobilization ... has significantly reduced the visibility of children in the forces as those who were among the first to be demobilized were those below 18, and women.[3]

There is no complete and reliable documentation about the number of killed, wounded, disappeared and displaced persons in four years of armed conflicts ... Therefore there is no such documentation about child soldiers either.[4]

The persons interviewed claimed that lots of minors did not show up to be listed by [the UN peacekeeping mission] due to lack of confidence and fear they felt in disarming "in front of the enemy".[5]

In the world at this moment over 300,000 children are participating in armed conflicts. It is impossible to give an exact figure as most of them are the "invisible soldiers" of our title.

They are invisible because those who employ them deny their existence. No record is kept of their numbers and ages, or the ages are falsified. Many are not part of the formal strength of the armed forces or armed groups to which they are attached, but are unacknowledged servants or hangers-on. Many are enrolled in a variety of militias and citizens' defence groups, the very existence of which is an under-documented area, rarely

---

[3] Case study for Ethiopia
[4] Case study for the former Yugoslavia
[5] Case study for El Salvador

included in official figures for armed forces. They are invisible because most spend their time in remote conflict zones away from public view and the prying lenses of the media. They are invisible because they tend to vanish. Many do not return from the battlefield because they are killed or, having been injured, are abandoned. When the conflict is over, they are rarely as visible among the demobilised troops as they were among the combatants at the height of hostilities. Additionally, those in their teens – the majority – are "invisible" because they are less obviously children – requiring a second look to realise that "they are not old enough to shave". And individually they all grow older. Within a few years even the child soldier who has survived has vanished, locked inside an adult soldier or an adult former soldier.

Of course the participation of children in armed conflict is not new. In the Middle Ages a boy who wished to become a knight would, from the age of about fourteen, serve as a squire, cleaning his master's armour and standing guard while his master slept. In the eighteenth and nineteenth centuries, "powder monkeys" as young as eight years old had the singularly hazardous task of loading the cannons on board ships, while on land infantry regiments used drummer boys for battlefield communication. The boy who lied about his age in order to enlist is a familiar motif in stories of the first and second world wars, both fact and fiction. At all times children have probably been employed as lookouts, spies and messengers.

There is a continuity from these historical examples to the experience of children in armed conflict today. They are still used wherever their small size and agility are seen as assets – for espionage, communications and demining – and from many countries come reports of recruits or younger children being used as the domestic servants and household guards of officers and their families.[6] Nevertheless, it is the period since the Second World War which has become the era of the child soldier as a result of changes in the nature of conflict so radical that not only have the numbers of children involved escalated, but the nature of their contribution has fundamentally altered.

The first such change is that "traditional" armed conflicts between states are now the exception rather than the rule. Depending on whose def-

[6]   Case studies for Burma/Myanmar, El Salvador, Liberia, Paraguay and Rwanda

inition is used, at any one time in recent years there have been anything up to fifty armed conflicts in the world, but in never more than two or three of these have the major protagonists been the armed forces of two sovereign states. The remainder are, whatever incidental international ramifications they may have, primarily "internal" armed conflicts, taking place within the home country of the combatants. This change in turn has inexorably led to a blurring of the distinction between the combatant and the "civilian" population. The proportion of "civilian" casualties has in consequence steadily risen: over 84% of those killed in armed conflict since the beginning of this decade have been civilians.

Most important, however, have been the changes in the weaponry used. Despite all the concern about the proliferation of weapons of mass destruction, 90% of casualties in recent armed conflicts have been caused by small arms. It is the development of lightweight automatic weapons, archetypically but by no means exclusively the Soviet AK47, which has taken the child soldier from the margins to the very heart of modern conflicts. The mediaeval squire could not hope to don his master's armour until he had reached physical maturity and even a generation ago battlefield weapons were heavy and cumbersome, limiting children's participation to support and ancillary roles but today there is no physical barrier to their serving from an early age as combatants on something like an equal footing with adults.

The changing nature of conflict has legal as well as practical implications. Although some peacetime recruitment of children is included, most of the situations studied in this book are non-international armed conflicts. In many of these, the government denies that it is involved in an internal armed conflict as classified by the 1949 Geneva Conventions or the 1977 Additional Protocols, and therefore denies the applicability of international humanitarian law. The consequences of this denial include lack of access by the International Committee of the Red Cross (although some governments do permit the ICRC to visit detainees while denying that there is an armed conflict), and that the government classifies opponents as criminals or terrorists (while in many cases freely killing members of armed opposition groups contrary to the right to life under international human rights law).

Yugos
Kosov

Atlantic Ocean

Mexico

Algeria

Sierra Leone

Liberia

C
B

Colombia

An

Peru

Child Soldiers

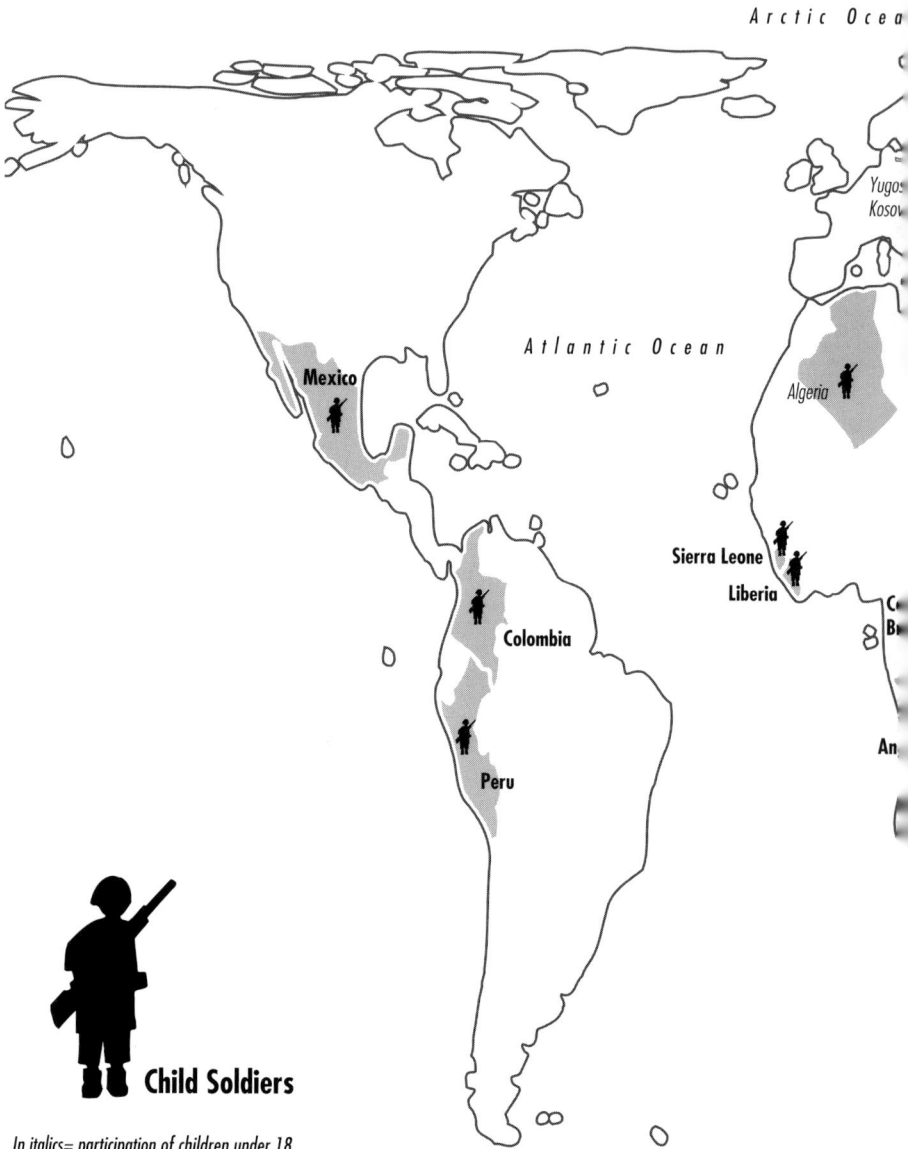

In italics= participation of children under 18
In bold letters= participation of children under 15

22

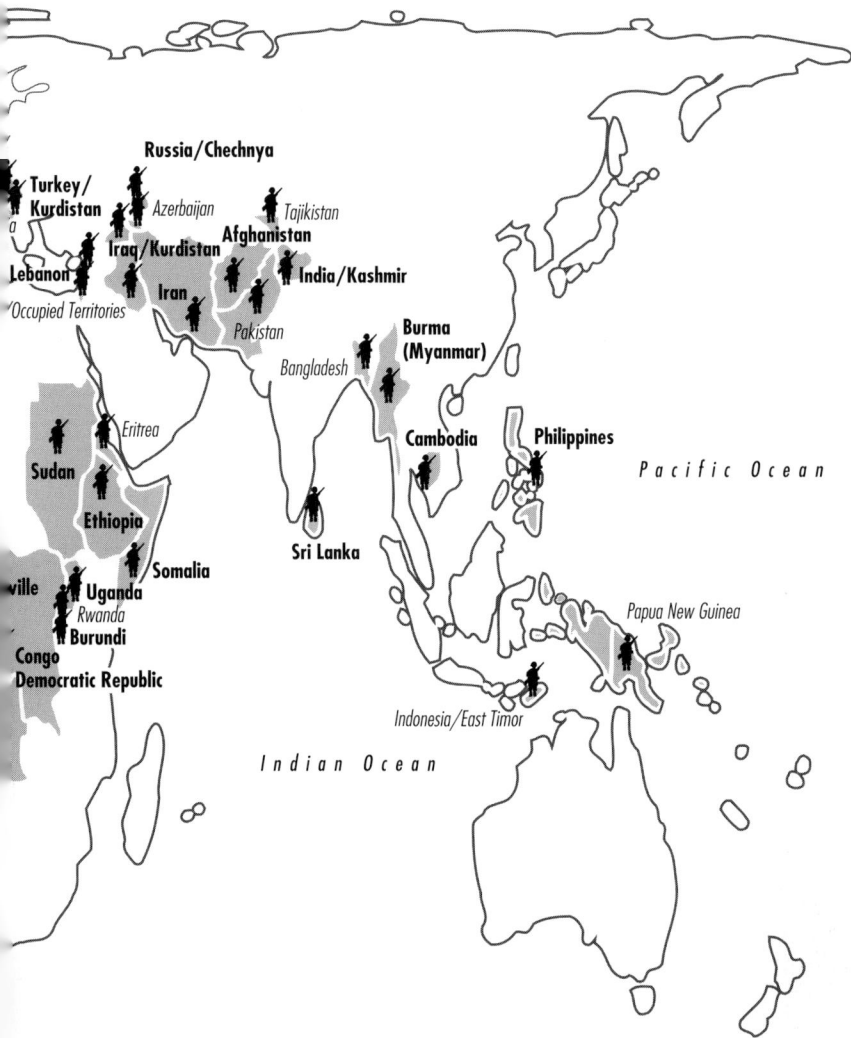

Russia/Chechnya
Turkey/Kurdistan
Azerbaijan
Tajikistan
Afghanistan
Iraq/Kurdistan
Lebanon
Iran
India/Kashmir
Occupied Territories
Pakistan
Burma (Myanmar)
Bangladesh
Eritrea
Cambodia
Philippines
Sudan
Ethiopia
Sri Lanka
Somalia
ville
Uganda
Rwanda
Burundi
Congo Democratic Republic
Papua New Guinea
Indonesia/East Timor

Pacific Ocean

Indian Ocean

Children under 18 years of age have been reported to participate in either government or opposition forces, or both, in armed conflicts in these locations (ongoing conflicts and conflicts in which hostilities ceased during 1997 or 1998).
This information has been compiled by Rädda Barnen.

Children under 18 years of age have been reported to participate in either government or opposition forces or both in armed conflicts in the following locations (ongoing conflicts and conflicts in which hostilities ceased during 1997 or 1998). Bold letters indicate the participation of children under 15 (28 cases). See map on pages 22–23 for a geographic overview.

Afghanistan
Albania
Algeria
Angola
Azerbaijan
Bangladesh
Burma (Myanmar)
Burundi
Cambodia
Colombia
Congo Brazzaville
**Congo Democratic Republic
  (former Zaire)**
Eritrea
Ethiopia
Indonesia/East Timor
India/Kashmir
Iran
Iraq/Kurdistan

Israel/Occupied Territories
Lebanon
Liberia
Mexico
Pakistan
Papua New Guinea
Peru
Philippines
Russia/Chechnya
Rwanda
Sierra Leone
Somalia
Sri Lanka
Sudan
Tajikistan
Turkey/Kurdistan
Uganda
Yugoslavia/Kosovo

This information has been compiled by Rädda Barnen.

Specifically, the effect on children is that they also may be treated as criminals or terrorists, subject to the same emergency legislation as adults with no account taken of their age in terms of treatment or punishment. Furthermore, the non-application of international humanitarian law means that it is harder to hold the armed opposition groups to any rules: if the government does not accept that humanitarian law applies why should they? This does not, however, absolve the government from its obligations under national law, any international human rights treaties to which it is a party (in particular in this context the Convention on the Rights of the Child) and customary international law. The apparent evidence of better treatment of recruits by armed opposition groups which invoke Additional Protocol II to the Geneva Conventions underlines the importance of governments becoming parties to the Protocol and of both governments and armed opposition groups being encouraged to accept its provisions.[7]

States parties to the Geneva Conventions and Protocols, and to the Convention on the Rights of the Child, are required to "respect and to ensure respect" for international humanitarian law.[8] Knowingly to recruit, or to allow other states or armed opposition groups to recruit, those who are likely to commit atrocities raises questions about the compliance of states with this obligation. Much of the testimony quoted in Chapter 4 leads to the conclusion that the greater suggestibility of children and the degree to which they can be normalised into violence means that child soldiers are more likely to commit atrocities than adults, given which it particularly behoves all states parties to the above treaties to consider their responsibility to prevent the recruitment and use of children in this way.

Many of these conflicts have their roots in poverty, economic and social injustice. At the same time, it is the poor themselves who pay the highest price – often conscripted or forcibly recruited into government armed for-

---

[7] See also F. J. Hampson, "Legal Protection Afforded to Children under International Humanitarian Law" (University of Essex, UK, May 1996), report prepared for the UN Study on the Impact of Armed Conflict on Children

[8] Article 1(1) of the four 1949 Geneva Conventions and Additional Protocol I, and Article 38(1) of the Convention on the Rights of the Child

ces, civil defence forces and armed opposition groups. They are unable to pay off recruiters, or are themselves paid by those better off to send their own children as substitutes. Those who join voluntarily often do so as one of the few ways to earn a living or support their families, or simply to survive. Without the possibility of education or employment, where else can they turn? The conflicts take place predominantly in poor rural areas, thus impoverishing them even more, making education and employment even less likely, destroying and displacing families and leaving the children even more vulnerable both to volunteering and to forcible recruitment. In the circumstances, is it surprising that children are often lured and intoxicated by the power and security of a uniform and a gun, and that this often leads them into a life of violent crime? This may be during the conflict itself or afterwards: reports include burglary, banditry, kidnapping, ambushing passers-by, armed robbery, theft, looting, rape, narcotic abuse and illegal gambling. Others, having learned to kill, graduate at a young age to being contract killers hiring themselves out to the highest bidder.

Moreover, the problems are not just of the individual child soldier or ex-soldier concerned. The greater the number of present and former child soldiers, and the younger they were at first recruitment, the more serious the implications for the whole society, recognising that many of them will leave the armed forces or armed groups as adults. A large part of a generation may have missed out on education, on the opportunity to acquire social and economically-useful skills. There is also the question of what they have gained in the course of their "service": many of them have suffered treatment that no civilised person could countenance and have learned the value of a gun as a means of livelihood and to get what they want. No society can afford not to take steps to address such a potential threat to its peace, security, stability and the rule of law.

NUMBERS AND DISTRIBUTION

Estimating of the number of children in the armed forces is a difficult task. We have looked at figures in casualty wards in hospitals; we have gone through a valuable resource of mine victims; we have looked at surveys on demobilisation; we have talked with staff of a major military training camp and we have discussed the numbers issue with anyone vaguely knowledgeable about the military and with all interviewees and the soldiers themselves. Not surprisingly, there is a wide variation in the estimates of the number of child soldiers involved.[9]

The total number of child soldiers in each country, let alone the global figure, is not only unknown but unknowable. Child soldiers are commonest in the thick of conflicts, precisely where accurate up-to-date information is hardest to obtain. As will be seen from Chapter 2, a large proportion are employed without enquiry about their age. Those who employ them, and who alone are in a position to hold definitive data on numbers and ages, rarely have any incentive to obtain or collate such data, and usually have a strong perceived interest in obstructing or deterring the access of third parties to such information. When those in authority are aware of existing national and international standards, they are reluctant to reveal breaches. This not only makes the collection of information harder, but may distort such information as is obtained.

Governments, armed opposition groups, the communities themselves and non-governmental organisations (NGOs) often deny the involvement of children even where the presence of child soldiers has been documented, or claim that these were isolated cases. In some instances, this made it impossible to obtain case studies. The fact of denial implies that, at least retrospectively, doubt is felt about the acceptance or encouragement of child participation. Unfortunately, its consequence is that information is more difficult to come by, thus making a greater understanding of the causes and consequences of such participation harder, as well as leaving individuals, their communities and their societies without help in addressing the experiences through which they have gone. Denial at the time has the further result that the situation of child soldiers is not considered or subject to debate or monitoring because their participation is unacknowledged, undocumented or falsified.

However, the accompanying map shows 36 current or recent[10] armed conflicts for which reports of the participation of children under 18 years of age have been received by the Rädda Barnen Project on Child Soldiers. Sixteen

[9]   Case study for Cambodia
[10] Conflicts in which hostilities ceased during 1997 or 1998.

Girls and boys are trained in Beirut 1986. Photo: Pica Pressphoto.

of the conflicts in question have been considered in the case studies which form the principal sources for the present book. Case studies were also undertaken for El Salvador, Guatemala, Mozambique, Nicaragua and the former Yugoslavia (ie Bosnia and Croatia), in all of which armed conflict effectively ended before 1997, and for Bhutan, Honduras and Paraguay, where no armed conflict has been taking place. (See pages 16–17 for an overview of the 26 case studies.).[11]

It should be noted that the case study on the Occupied Territories deals only with the Intifada – the involvement of children in unstructured and essentially unarmed acts of politically-motivated violence – and that similar situations are mentioned in the case studies on Northern Ireland and South Africa. While at times such events have been discreetly orchestrated and co-ordinated by armed opposition groups, there is no doubt that in all three places they also occurred spontaneously, in which instances the participants could not be regarded under even the broadest of definitions as "child soldiers".

The best guess which can be made – on the information available in May 1998 – of the number of children recently involved in the thirty-six conflicts where their participation has been documented is over 300,000.[12] That this is larger than the figure of 200,000 which was suggested when the issue of child soldiers first attracted attention in the 1980s (and which has been the ultimate source for almost all subsequent attempts to produce an overall total) is not necessarily because of a dramatic increase in the last decade but simply of better information. It must also be observed that much of what has been published regarding the involvement of children in particular conflicts has explicitly or implicitly referred not to those under the age of eighteen, which is the general age of majority contained in Article 1 of the 1989 Convention on the Rights of the Child (CRC) but to

---

[11] Bhutan has been classified as one of 40 "serious dispute situations" falling short of open armed conflict by A. J. Jongman & A. P. Schmid, "World Conflict Map, 1994–1995", PIOOM Newsletter and Progress Report, 7, 1 (Leiden University, 1995), p 23. The case study for Honduras documents the use of children by armed opposition groups.

[12] For a global survey of numbers and other relevant information see Annex 1.

the lower age limit of fifteen established in Article 38 of that Convention as the minimum age for recruitment into armed forces and for participation in hostilities (which is also the age stated in the 1977 Additional Protocols to the 1949 Geneva Conventions governing the conduct of armed conflicts). Indeed, if all recruitment of under-18s is taken into account, including that undertaken in "peacetime" and in conformity with the applicable national legislation (and the case for doing so will be made in Chapter 7), then the numbers world-wide would be more than doubled.

But at the end of the day, all speculation about precise numbers is barren. Almost by definition information on child soldiers is out of date before it is published. On the one hand, usually it is only when political or military developments start to transform the situation in a country that substantial information on the past existence of child soldiers comes to light. On the other hand, new conflicts continue to erupt and it takes time for details such as the actual ages of the participants to be ascertained, let alone reported, particularly as the involvement of children tends to occur and increase as the conflict progresses rather than characterising the initial stages. In any case, the individual children themselves are always growing older, thus the most accurate statistical information remains relevant only if the process of child recruitment continues unchanged. Even if all the obstacles to obtaining information could be surmounted, the facts themselves would still remain elusive: of the former Renamo child soldiers analysed in the case study on Mozambique, 16% of the boys and no less than 28% of the girls did not know their own ages.

The "snapshot" approach falls into the trap of thinking of a child soldier as a static phenomenon. In fact the individual child soldier is essentially transitory. Within a few years, he – or she (see the section on "Gender" in Chapter 3) – will be an adult soldier, a former soldier, or dead. Each individual child soldier is youngest at the point of recruitment, but where does the experience end? The model which would have the typical child soldier going from recruitment through service to demobilisation is far too simplistic. As Chapter 5 makes clear, the demobilisation of child soldiers is rarely routine, but usually takes place as part of a mass demobilisation following a peace accord. (Sadly, the demobilisation often proves to be as temporary as the peace itself.) For some, of course, the experience ends in

death; for others the end may come through injury, desertion, or being captured (which may however simply return the child to the conflict on the
opposite side). There are rare instances where not only is a fixed-term period of conscript service adhered to in practice, but where some conscripts
are so young that they have completed this term of service before they
reach their eighteenth birthday. In general, however, and in the absence of
dramatic political developments, it is rare for a child soldier to come to the
end of his or her fighting life as a child. From this it may be deduced that
almost without exception the numbers of child soldiers in any particular
force at any particular time will increase with age: thus by far the largest
number of under-18s in the force will be aged seventeen. Many, however,
will have been recruited at a far younger age:

> ... the first soldier demobilised by [the armed opposition group] was 16.
> The same boy relates that he was recruited at the age of 9[13]

Details of recruitment ages are even harder to come by in a systematic form
than the current age distribution of child soldiers. However what evidence
there is indicates that there are two significant thresholds, one just before
the teens – that is, at about age 10 to 12 – and one at 15. Although the case
studies and other sources contain frequent references to very young child
soldiers, such extreme figures are unlikely to be representative, serving
simply to draw attention to the full size – and horror – of the problem. (In
Colombia, Lebanon, Liberia, Mozambique, Peru and Uganda children of
less than 10 are reported to have fought in opposition armed groups or the
government-armed militias which often form the front line against them.)
Even where there is clear evidence of the recruitment of children aged in
single figures, this seems to represent a very small proportion of total
recruitment. There is a very sharp increase in numbers from somewhere
around the age of 10 – the age at which the average child is just big enough
to carry and operate an AK47. Where there are figures for the recruitment
of children into "regular" government armed forces and even where forced
under-age recruitment is rife, those in their early teens or younger tend to
be the exception; there is a sharp rise in numbers over the age of 15, and

---

[13] Case study för Mozambique.

in general the incidence of recruitment increases right up to the legal (national) recruitment age. The result of these different trends in recruitment is that, although the overwhelming majority of all child soldiers are to be found in government armed forces, those in the early teens and younger are more likely to be found in armed opposition groups.

There is no direct relationship between recruitment ages and the proportion of a given force which is composed of under-18s. In an army largely composed of short term conscripts, the average age will be close to the typical recruitment age. Such armies were reported in more than one case study to have more than half their strength aged under 18, with the proportion rising as high as 80%. By contrast, a professional volunteer army may attract almost half its recruits at under the age of 18 and yet have this age group forming a very small percentage of its total personnel because of the longer duration of service. Most armed opposition groups would appear to fall somewhere between these two extremes. Although many rely heavily on the recruitment of children, service is usually for the duration of the conflict and the estimates of the overall proportion of children in the groups typically show them outnumbered by two or three to one by adult members, that is, children form some 25%–40% of the total.

These figures, of course, have their converse:

...literally thousands of youths would have been involved in such ... situations over the last 25 years [14]

As a conflict into which children are recruited continues, so the number of adult soldiers who started their involvement as children will rise, and can rapidly become a majority. There will be a similar cumulative effect on the entire society. To illustrate this, one might take the civil war in El Salvador. From about 1982 to 1990 it was reported that the government was recruiting between 12,000 and 20,000 youths annually, almost all below the official age of 18. These recruitment figures would imply that in a country of 5.5 million, there have been well over 100,000 who participated in the war as child soldiers on the government side alone. How many have survived? Where are they now? What have been the physical and mental health

[14] Case study for Northern Ireland

impacts? Indeed, how many of the age group affected avoided the fate of becoming a child soldier on one side or the other? (*Who* avoids service is addressed in Chapter 2.) Hardly any research, anywhere, has yet been done on questions like these.

Not only are the effects cumulative: there is evidence that the use of child soldiers, particularly in conflict situations, rarely reaches a state of equilibrium. The situation tends to develop over time. This development in some rare instances takes the form of an end to child recruitment, which very rapidly leads to the disappearance of child soldiers (though not, of course, of former child soldiers). Far more often, however, the pattern is that children are not very much – or at all – in evidence in the early stages of the conflict but that as it continues more, and younger, children are gradually drawn in.

Several of the case studies report or imply that the problem is a growing one. Burma/Myanmar has seen a series of conflicts ever since the 1949 war of independence, but all of the former child soldiers interviewed for the case study had joined the conflict since the 1988 student uprising, and some reported that it was only after 1988 that widespread recruitment of children by government armed forces began. In Guatemala, the armed conflict dates back to the 1960s, but the foundation of the PAC militias, the heaviest users of child soldiers, did not occur until the early 1980s. The case study for Cambodia reports no evidence of the recruitment of significant numbers of children in the 1960s, and dates the emergence of the phenomenon to the "Lon Nol period" (1970–1975). Child soldiers have featured prominently in all subsequent stages of the Cambodian conflicts. The case study for Nicaragua (the earliest in terms of the events with which it deals) analyses a group of child recruits to the National Guard captured following the defeat of the Somoza government in the civil war of 1976–9; it gives no indication of the overall scale of juvenile recruitment at that date, but it is known that of 22,000 irregular combatants demobilised after a further ten years of conflict in Nicaragua, 1,800 were less than 16 years old.[15] It would

[15] Source: UNICEF 1992, quoted in J. M. Tortorici, "Peace Education in Nicaragua" in M. McCallin (Ed.): *The Psychological Well-Being of Refugee Children. Research, Practice and Policy Issues* (ICCB, Geneva, 2nd edition 1996)

not appear that the problem had declined in the intervening period.

It is, however, the case studies on Ethiopia and Afghanistan which provide the most thought-provoking evidence on developments over time. In Ethiopia, the 1974 Marxist coup was followed by the growth of a number of political and regional armed opposition groups which mobilised a broad spectrum of the population, including children. Major recruitment drives were held by the government in 1977, 1982 and 1990. The degree of compulsion and the comparative youth of those targetted increased each time. By the time the government fell in 1991, perhaps 15% of its troops were under 18, while the proportion of children in the victorious opposition forces may have been as high as a quarter. From both sides it was the youngest soldiers and the most recent recruits – usually the same thing – who were demobilised most hastily: thus the evidence vanished before they could be counted. The victory of one opposition faction did not herald the end of the conflict, which has continued on a number of fronts. The most recent event reported in the case study is the defeat in 1992 of one faction which had split from the governing coalition, resulting in the capture of 22,000 opposition troops of whom 8,490, or just under 39%, were aged under 18.

The case study for Afghanistan reports that after the outbreak of the civil war in 1978 the official recruitment age was reduced from 22 to 20, to compensate for defections to the ranks of the Mujahideen, and eventually to 18. By the time the Najibullah regime fell in 1992, according to the nationwide survey conducted for the case study, some 26% of its force was aged under 18, including 10% aged under 16. A very slightly higher proportion of the opposition Mujahideen forces were aged under 18. As in Ethiopia, the victorious opposition promptly split into various factions and the fighting continued unabated. The case study estimates that no less than 45% of the strength of these factions was aged under 18, and furthermore, that the proportion of front-line fighters among the children had increased to two-thirds.

In all of these cases it appears that the warring parties, having first soaked up the men of what would normally be considered military age, found themselves turning to ever-younger children in order to fill their ranks. There is also reason to fear that a vicious circle of pre-emptive recruitment developed, with each side trying to enlist youngsters before they could be recruited by the other.

By contrast, instances where the number of child soldiers has unequivocally declined are rare. High profile demobilisations, as in Liberia and Sierra Leone, are often just one element in complicated political manoeuvrings behind which the dynamics of conflict and of the employment of child soldiers in due course prove to have been dormant rather than extinct. The case studies for El Salvador and Mozambique do report situations where armed conflicts have clearly been brought to an end and the demobilised child soldiers have not been simply replaced by new recruits. The government successes against the Shining Path in Peru have so depleted the numbers of that group that there must have been a substantial net decrease in the number of child soldiers in that country.

Above all, the change of government in South Africa has meant that conscription, which used to take place at the age of 16 (for white males), has been abolished in favour of an all-volunteer force with a minimum recruitment age of 17. Despite the amalgamation of the former government and opposition forces, this has led through attrition and simple ageing to a rapid reduction to a minimal number of child soldiers. However, this progress has not been achieved without a disturbing parallel development detailed in the case study, namely the large number of young people who have, apparently, organised themselves into "Self-Defence Units" and "Self-Protection Units, which are armed but not treated as armed opposition groups, have an ambiguous status vis-à-vis the factions formerly engaged in the war against apartheid (and whose rivalry continues in parts of the country to erupt in incidents of open violence), but are not answerable to civil authority. The case study for Colombia mentions similar "popular militias", which claim to be community-based but whose allegiance is uncertain.[16]

Analysing the trends, the case study for Afghanistan points to a strong association between, on the one hand, the increasing number and decreasing age of the children involved in the conflict and, on the other, the destruction of the economy, but above all the almost complete loss of educational facilities. It estimates that some 90% of the country's children have no access to schooling. Furthermore:

[16] However, the case writer comments, "at present some sectors of the militias may have relations with guerilla groups".

The continuous involvement of child soldiers in war can be a cause of further deterioration in the security situation in future. This may even result in a situation that, in reality, peace will not show its face for decades.[17]

In other words, the extensive involvement of children as combatants may in itself be a significant factor in prolonging the conflict.

An increasing number of factions with an ever-younger armed membership, the development of overwhelmingly juvenile armed groups whose claim to legitimacy is unclear, children growing up without education, trained in combat, and deriving their meaning in life from their participation in the conflict in one form or another, this is the situation which now confronts many of the countries considered in the case studies. Of what future might it contain the seeds, if allowed to develop unchecked?

[17] Case study for Afghanistan

Boy soldiers in guerilla training camp, El Salvador.
Photo: Martin Adler, Panos Pictures.

CHAPTER 2
Recruitment

It is definitively a problem of political ethics, the recruitment laws themselves establish the age but if states themselves allow the armed forces not to respect the law we shall never get anywhere.[18]

It is no accident that such a heavy emphasis in this book is placed on recruitment. Recruitment is the defining moment. Without the recruitment of children, there would be no child soldiers. Many child soldiers are invisible because they grow into, and become demobilised as, adult soldiers.

In drafting the questionnaires, reference was made to compulsory, forced, voluntary and induced[19] recruitment, and the writers of the case studies took pains to relate their answers to these categories. What emerges, however, is that the areas of overlap between the categories are more striking than the differences. In practice there would appear to be a continuum between forced and voluntary recruitment, with the true facts in many cases being unascertainable.

CONSCRIPTION

Of all the categories, that of compulsory recruitment, or conscription, is the most distinct. This consists of the legal obligation of citizens falling into the specified category to perform a stated period of obligatory military

---

[18] Dr Ramon Custodio, President of the Committee for the Defence of Human Rights in Honduras, quoted in the case study for Honduras (translated from the Spanish).

[19] Induced recruitment was defined as those situations where there is no proof of direct threat or intimidation but the circumstances suggest that the enrolment is not voluntary.

service. Conscription is practised in many countries in all regions of the world. In some it has been introduced as a response to conflict, in others it is a standing arrangement. In some countries, what are referred to in this book as "child soldiers", that is, those under 18, may be legally conscripted. In the majority of cases, however, military service is required of males (and sometimes females) of 18 years and over (see Annex 1 for the current position).

It is notable that none of the case studies mentions a possibility of exemption from military service on grounds of conscience, or the availability of alternative civilian service for those with a conscientious objection to military service.

By its nature, conscription is a governmental prerogative. In one instance, the armed opposition group issued its own "Compulsory Military Service Law", aimed at members of the ethnic minority of military age (18–25 years).[20] However, in reality this was simply a way of encouraging volunteers and of legitimising forced recruitment to pre-empt conscription into the government armed forces. In fairness, it should be observed that the same applies to much that is done by governments in the name of conscription. For, unfortunately, the existence of conscription with a legal minimum age of eighteen is no guarantee against the recruitment of child soldiers into government forces. The main reasons for this, which will be discussed below, are: lack of documentation, voluntary enlistment for compulsory military service, "quota" enlistment, and what may be termed legal, systematic and infrastructural inadequacies.

Lack of Documentation: The simplest, and the most intractable, problem is lack of birth or identity records. The absence of these leads to genuine problems in establishing the age of potential recruits, as well as serving to facilitate under-age recruitment:

> *It is fairly usual for age to be estimated based on stature, psycho-motor development, and, in the case of girls, the appearance of breasts.*[21]

---

[20] Case study for Turkey
[21] Case study for Mozambique (translated from the Portuguese)

42

Even where documentation does exist, there are circumstances where it commonly overstates the true age of the child:

> In the case in which the elder brother dies, of disease or accident, at an early age, to avoid the painfully long process of issuing state identification cards for the younger brother, some families have preferred to transfer the identity of the deceased elder to the younger. As a result, early recruitment has taken place.[22]

Or it may even represent a *post hoc* rationalisation:

> For the majority ... especially for those living in rural areas precision in the year, let alone in the month and day of birth, is an unattainable abstraction. Hence age is determined not by having regard to any official record but by having regard to what the person says his/her age is, physical appearances, or even by what the person seeking the information thinks the age is. In relation to children's involvement in armed conflicts and armed forces, this situation means that those below 18 are officially recorded as above 18 simply because the recruitment of children is officially not acceptable and there is no way of ascertaining the actual age of the persons concerned.[23]

This uncertainty, credible in itself, may readily be used either as an excuse for failure to make such checks of age as are possible, or to mask conscious and systematic under-age recruitment. Those forcibly recruited or volunteering are encouraged or forced to state that they are 18 in order to ensure apparent conformity with national legislation or international norms. When questions are asked, the authorities blame the recruits for misstating their ages. Alternatively, "informal lists" may be kept of under-age soldiers rather than including them in official records.[24]

---

[22] Case study for Turkey

[23] Case study for Ethiopia

[24] Case studies for Burma/Myanmar, El Salvador, Ethiopia, Guatemala, Mozambique and Paraguay; see also Human Rights Watch/Africa, Children's Rights Project: *Children of Sudan* (September 1995), p 26

Young recruit in the army of Paraguay. Photo: Jorge Sáenz.

Voluntary Enlistment for Compulsory Military Service: Sometimes, faced with the inevitability of military service, recruits find it more convenient to enlist before the official age. In many cases voluntary enlistment is permitted at a lower age than that of conscription. Sometimes there is specific provision for voluntary induction in order to complete compulsory military service at an earlier and more convenient time. These loopholes may be exploited to mask systematic recruitment of children, with underage conscripts being claimed as "volunteers". In other circumstances, instances have been documented of under-age candidates presenting themselves with forged documents for recruitment.[25]

"Quota" Enlistment: Not all conscription systems rely upon an individual call-up. Indeed, such a practice can only function where there is relatively accurate and comprehensive documentation of the population. Often it is the number rather than the identities of the conscripts which are predetermined. One method of filling the quota is the conscription lottery which takes different forms in different countries. One case study described a system by which all those deemed eligible on the basis of height ("A rifle was used as the measure of appropriate height!") had to draw out a piece of paper, some of which were blank and some said "join the army". However, the apparently random nature of the process was belied by evidence that the lottery was rigged in advance so that certain families or individuals were targeted, in particular those who had fallen foul of the recruiter.[26]

If those entrusted with producing the recruits are unable, for whatever reason, to come up with the requisite number of the official recruitment age, they may be tempted, or forced, to make up the shortfall with underage recruits:

Personnel of the armed forces entrusted with the task of recruitment were officially under a duty to exclude children from recruitment. They did not live up to expectations, however, due to several factors. The

[25] Case studies for El Salvador and Turkey
[26] Case study for Cambodia

major reason for this is that the main task of these personnel is to obtain
a pre-set number of recruits within a specified period of time. Since not
so many volunteered to join the army, the recruiters take as many of
those as present themselves unless their recruitment is blatantly not con-
forming to the criteria for recruitment including the age requirement.[27]

These agents may be detachments of the military. On the other hand, the
task of finding recruits may be devolved to local militias/civil defence forces, or the headmen or other authorities of the local community:

> *The district/sub-district Administrator ... instructs the village headman*
> *to provide a specific number of people from his village for "voluntary"*
> *recruitment, which he is obliged to meet. Thus several (mostly) young*
> *people from the district are rounded up in district headquarters for*
> *initial medical and other check-ups by authorised Armed force person-*
> *nel.*[28]

Such indirect methods of conscription are particularly prone to corruption
and substitution, both of which increase the number of children recruited.
The selection may in practice be influenced less by age than by the inability to pay bribes, whether to the recruiter or to a potential substitute, having had a disagreement with the village leader or for example, a younger
brother may be put forward in order to allow his older brother to finish his
education.

A quota system will often coexist with the requirement that only a
proportion of the eligible population need perform military service,
although neither of these conditions relies upon the other. Where the two
coincide there are implications not only for the likelihood of the recruitment of child soldiers but also on who is vulnerable to such recruitment, as
discussed in Chapter 3. In such circumstances, there is in practice little
difference between conscription and forced recruitment. Indeed, armed
opposition groups sometimes themselves obtain recruits by imposing quo-

[27] Case study for Ethiopia
[28] Case study for Bhutan

47

tas upon communities within their area of operation; sometimes the recruits go willingly, sometimes not.[29]

Legal, Systematic and Infrastructural Inadequacies: Where the conscription system is fundamentally flawed, rather than simply subject to abuse at the margins, conscription frequently merges into forced recruitment, particularly if it can be done away from public scrutiny, because the conflict is taking place in a remote area, for example, or if disenfranchised sections of the community can be targeted.

The constitutional provision for conscription may not be translated by legislation into appropriate procedures, so that in practice implementation is at best random. Alternatively, the military may simply ignore the prescribed procedures. Even where laws are in place to punish those engaging in under-age recruitment, these may never be applied, and thus be ineffective.[30]

The system or infrastructure may be inadequate to produce the required number of recruits particularly if, as a result of conflict, large numbers are required over a prolonged period of time, or if the mechanisms for enforcement are so inadequate that a high proportion of those eligible escape service and alternative recruits have to be found. Evasion is particularly common where the conflict, or the military, are unpopular, or where casualties are known to be high.[31] If salaries are inadequate and desertions high, "ghost soldiers" may be maintained on the lists in order to provide the commanders with surplus salaries while children are recruited to fill the ranks.[32] On the other hand, as will be discussed below, there may be a deliberate policy of the military to target certain groups and the political structures may lack the strength, or the desire, to challenge it.

---

[29] Case studies for Burma/Myanmar, Cambodia, Lebanon and Liberia
[30] Case studies for El Salvador, Honduras and Paraguay
[31] Case studies for Afghanistan, Cambodia, El Salvador, Ethiopia, Guatemala and Paraguay; see also Human Rights Watch/Africa, Children's Rights Project: *Children of Sudan* (September 1995), p 55
[32] Case study for Cambodia

Causes: Forced recruitment into government armed forces may be a response to an immediate shortfall in manpower,[33] as discussed above. However, in some countries there is a long tradition of forced recruitment.

In addition to the generally disenfranchised, there are particular groups which are especially subject to forced recruitment. These may be specific ethnic, racial or religious groups, such as the indigenous population, because they are perceived as a threat to the government. This threat may be as a recruiting ground for armed opposition groups, or their general supporters or potential supporters of it, or merely as the "sea" in which the opposition groups "swim" (in Mao's phrase). Such recruitment frequently forms part of a general campaign of intimidation, repression or disempowerment and disruption.[34]

As already mentioned, forced recruitment into armed opposition groups may sometimes be justified as a form of conscription, a requirement that all members of the ethnic group contribute to the armed struggle.[35] Otherwise much the same reasons apply as in the case of government armed forces: the group is under pressure and needs additional soldiers, which is particularly likely if the conflict is prolonged, and is compounded if the group is unpopular and therefore the number of volunteers is inadequate, or if there is a high level of desertions.[36] Whereas the government can, at least theoretically, use legal means to draw on the entire population, armed opposition groups can recruit actively only in the territory they con-

---

[33] Case studies for Afghanistan, Bhutan, Burma/Myanmar, Cambodia, Guatemala, Lebanon, Nicaragua, Paraguay and the former Yugoslavia; see also Human Rights Watch/Africa, Children's Rights Project: *Children of Sudan* (September 1995), p 55

[34] Case studies for Burma/Myanmar, El Salvador, Ethiopia, Guatemala, Mozambique, Nicaragua, Paraguay and Peru

[35] Case studies for Burundi, Lebanon and Turkey

[36] Case studies for Afghanistan, Burma/Myanmar, Cambodia, Lebanon, Liberia, Mozambique, Turkey and Uganda; see also Human Rights Watch/Africa, Children's Rights Project: *Children of Sudan* (September 1995), p 77

trol or have access to, thus at best they have a limited pool of manpower. One case study referred to the problem being compounded by a high level of migratory labour, causing a local shortage of able-bodied men.[37]

Methods of Forced Recruitment: The most common method of forced recruitment by both government and armed opposition groups is that described in the case study for Ethiopia:

> *An extreme version of forcible recruitment is the press ganging. This is known as "Afesa" in Amharic. A group of armed militia, police, or party cadres would roam the streets and marketplaces, picking up any individual or rounding up any groups they came across. Alternatively, they would surround an area and force every man and boy to sit down or stand up against a wall, using a threat of opening fire. All those eligible would then be forced on to a truck and driven away. Young men were recruited while playing football, on side streets and alleyways, going to school or market places or attending religious festivals. Teenage boys who worked in the informal sector selling cigarettes, matches, sweets, chewing gum and lottery tickets were a particular target.[38]*

Very similar accounts occur in different cultures:

> *The forced recruitment took place in the poor suburbs, at the football grounds, movie theatres, at the bus stops or in front of big schools, etc; places where young men and boys of the lower classes of society moved in major quantities. Army trucks also went to poor distant villages in the countryside and swept up the streets picking up the young men.[39]*

> *Other children were recruited at school gates, market places and other areas where people gather together.[40]*

> *"We were leaving school at the end of the day and the ... soldiers sur-*

---

37 Case study for Mozambique
38 Case study for Ethiopia
39 Case study for El Salvador
40 Case study for Mozambique (translated from the Portuguese)

*rounded the school ... There were 40 or 50 of us all leaving together, and
we were all arrested. We were all 15, 16, 17 years old, and we were all
afraid of the soldiers. We were students, we looked like students, because
we were all wearing our blue shirts and green longyis. Our teachers all
ran away in fear. Everything was in chaos ... We were all terrified, but
we couldn't even call out to them to let us go and that we were under
18, because we were so scared ... I didn't know what was going on and
they didn't explain anything to us."*[41]

In rural areas, both government and armed opposition groups use similar
methods. The army or group enters villages and small towns, killing people,
abducting children, looting and burning homes. Children are abducted
from their homes at night or during the day from home or school or from
working in the fields. In almost all instances they are taken by force or with
threats and intimidation:[42]

*"They take the boys and men as porters, rape some of the women, steal
the chickens and pigs, steal the rice and burn what we have in the
fields."*[43]

*Those who resist would be cut with pangas. Quite a number of victims
had their lips and ears chopped off in macabre rituals.*[44]

Various methods may be used to try to escape such recruitment; the cost
of failure can be high:

*If there was an "afesa" in a neighbourhood, local women would patrol
near the area and warn men and boys to stay away, or give them
"gabis" to hide under to disguise themselves as women. If all failed, people would run for hiding places into nearby houses, granaries, ward-*

---

[41] Case study for Burma/Myanmar
[42] Case studies for Burma/Myanmar, Cambodia, Colombia, El Salvador, Liberia,
Peru and Uganda; see also Human Rights Watch/Africa, Children's Rights
Project: *Children of Sudan* (September 1995), p 56, and Amnesty International,
"Sierra Leone: Human rights abuses in a war against civilians", 13 September
1995, pp 19 and 23
[43] Case study for Burma/Myanmar
[44] Case study for Uganda

robes, ceilings, cattle fodder, empty barrels, etc. There were also numerous instances of people trying to resist or escape press-ganging being summarily killed.[45]

At the margins, "press-ganging" by government forces can be described as an enforcement of conscription legislation. There is evidence that in some situations there is enough respect for age limits that those who can produce documentary proof that they are under-age are released, even if this tends to be associated with a preference for concentrating upon those groups who are unlikely to have such proof.[46] More frequently, age is at best a matter of indifference to those recruiting in this way; they are often satisfied with obviously under-age recruits:

> *Many children whose age has been mentioned clearly in their national ID cards as less than 18 years were taken to [a] special military commission where the military officers amended their age to meet the criteria of military service. In this way they were sending children aged less than 14 years to the armed forces.*[47]

There is reason to believe that children are disproportionately vulnerable to forced recruitment procedures. Recruiters prefer to concentrate on those who can resist least effectively, which includes children.

Sometimes, however, the evidence is that the recruitment of children is a deliberate policy rather than a consequence of other factors. The frequent references to recruitment in or near schools[48] point to this conclusion. Recruitment from schools is one instance in which those who are the most disadvantaged may not be the target, depending on how widespread the availability of schooling is. One specific instance was reported of a focus on national educational institutes to produce more educated recruits in order to combat urban armed opposition groups.[49]

---

[45] Case study for Ethiopia
[46] Case studies for Colombia, Ethiopia and Guatemala
[47] Case study for Afghanistan
[48] Case studies for Afghanistan, Bhutan, Burma/Myanmar, El Salvador, Ethiopia and Mozambique
[49] Case study for El Salvador

As previously mentioned, a frequent expedient of both government and armed opposition groups is to conduct their recruitment by proxy. Recruitment is entrusted to local authorities or militias, who have to fill a quota with the penalty for failure being to be sent to prison or on active service themselves.[50] This frequently degenerates into forced recruitment:

In an over-zealousness to fulfil the quota and to avoid the much dreaded active military service, the militias and military commissariats resorted to forcible and arbitrary recruitment. Trying to avoid the ill-will of their neighbourhoods, territorial militias in many cases picked up recruits from among strangers in market places, places of worship ... and on streets. They were not, however, sparing of their neighbours also, and sent letters of recruitment, and followed these with night time searches.[51]

However, even under age, specific individuals may be sought and, if they are not found, pressure may be brought to bear on their families. A vivid instance is provided by the following testimony of a former child soldier:

"I was 16 years old ... while my younger brother was 14 years old ... One afternoon a neighbour came to our home and told my father that the people's militia have a plan to visit our home during the night for the purpose of recruiting either myself or my younger brother ... my father called myself and my younger brother and told us that he will be taking us to a far away hiding place ... On the third or the fourth day, my younger brother urged me that he should go back home and find out what has happened ... When he reached home, he found out that our house was searched at the said night and the people's militia trying to find either of us, have detained both of our parents as hostages. He went straight to the police station, where they were detained, submitted himself for recruitment, and demanded the release of our parents. Upon their release, my parents came to take me home because they have got an assurance that not more than one person will be recruited from a single family. When I learned the fate of my brother I felt like crying. It

[50] Case studies for Burma/Myanmar and Ethiopia
[51] Case study for Ethiopia

felt as if I was guilty of betraying my younger brother. What is more, I could not bear the sight of my mother and sisters crying, and the thought of my younger brother being recruited to spare me, made me restless. In short I hated myself ... In the evening I somehow managed to sneak out of the house and went to the recruiting officials with the intention of substituting my younger brother. They did not show him to me then. It was several months later, at the training camp, that I learned my brother was not released and was in fact placed in a different training centre. He died a year later while in combat ..." [52]

The various forms of local militia (which are known in different places by such titles as civil defence patrols, self-defence committees, peasant patrols or temporary village guards) are an under-documented phenomenon whose role in recruiting children for themselves and for regular government armed forces merits further research. [53] Characteristically justified by governments because of the security situation in conflict areas, persons recruited into them are often under threat of otherwise being accused of being guerillas or sympathisers, with dire consequences for the individuals or the village concerned. Children, in particular, often serve on behalf of their parents, other economically active adults, or old or infirm people who are incapable of carrying out the duties required. Some militias may be genuine self-defence groups organised by young people from the local community in the face of inadequate protection from, or violations by, the government, but there are questions about possible links to one or other party. Alternatively, what began as local "peaceful, democratic and autonomous" organisations can be taken over by the government and armed, or permitted to carry arms, as part of its anti-subversion strategy. Where such forces are well paid and are created in an area of high unemployment, this attracts volunteers, but also creates employment which is entirely dependent on the continuation of the conflict. [54]

---

[52] Case study for Ethiopia

[53] Case studies for Cambodia, Colombia, Guatemala, Mozambique, Peru and Turkey; see also Human Rights Watch/Africa, Children's Rights Project: *Children of Sudan* (September 1995), pp 31, 57

[54] Case studies for Colombia, Guatemala, Mozambique, Peru and Turkey

Recruitment by intimidation rather than by crude compulsion is certainly not a practice confined to government forces. Even where armed opposition groups may genuinely believe that they are using persuasion rather than force, the children may feel that they have been recruited against their will.[55] Also, when a heavily armed group enters a village and makes speeches calling for volunteers, how voluntary is the recruitment? What are the perceived prospects of resisting?

On the other hand, if supplying volunteers to the armed opposition group bears the risk of retribution by government forces, there may be a systematic practice of reporting them as having been abducted. This may also be seen as affording some measure of protection to the individuals themselves should they fall into government hands:

What was common knowledge even to local officials at that time was, however, that although the [opposition] recruitment activities were reported as kidnapping by the national press ... this was mainly the result of the fear shared by the villagers of possible [government] retribution to suspected voluntary participation ... It was definitely in the interest of the family and the person taken by the [opposition group] to be reported as kidnapped in any case. First, he or she may have been truly abducted, but families would fear [government] interception of the specific guerilla group and attempt to protect their relatives by emphasizing the point of having been kidnapped. Then there was the possibility that the individual would volunteer to join the [opposition group] but this would leave the family with the burden of explaining his sudden disappearance to local officials – especially if the said person is obliged to serve the [government] army. Even before being trained by the guerillas they could be intercepted by [government] forces and caught. There was thus fear that if the potential recruits returned to their villages in one way or the other, before or after training, they could be punished by the authorities if suspected of voluntary recruitment. As a result of this, most voluntary recruitment was also reported by the villagers themselves to local authorities as kidnap. There were also cases ... during

[55] Case study for El Salvador; see also Amnesty International, "Sierra Leone: Human rights abuses in a war against civilians" (13 September 1995), p 23

which the individuals joined the [opposition group] of their own free will
and after initial military training returned to their settlements to con-
duct undercover military activities. Such people were explained off as
having managed to "escape" from the [opposition group] or "released" at
one stage. Another frequent explanation behind kidnap claims, as
reported by families and militants concerned, appeared to be concern on
the part of the volunteer or the family that the individual would not fit
rural war conditions and would be turned down half-way by the gueril-
la unit escorting them ... Yet there are many recorded incidents of [oppo-
sition group] abductions in the history of the conflict ...[56]

In short, the truth about individual cases may be impossible to ascertain in
the context of the conflicting pressures.

Less directly, some governments and armed opposition groups are
known to use the creation of educational establishments and institutions
for street children and orphans as providing training grounds for the mil-
tary, "warehousing" them against future need.[57] Those enrolled in such
establishments may include children captured from "enemy" positions or
abandoned in front line villages:

... orphans between 0–12 years, were taken by [the opposition group]
into refugee camps and daycare centres, where military training was
sometimes secretly given by [opposition] combatants ... many of those
children joined [the opposition group] later, at the age of about 12
years.[58]

Such institutions may be an attempt to respond to a genuine problem of
what to do with unaccompanied children. However, they also play on the
lack of security and educational opportunity for children. At best they
undermine the child's free choice as to whether or not to volunteer. At
worst they are simply a disguised form of forced recruitment.

[56] Case study for Turkey
[57] Case studies for Burma/Myanmar, Ethiopia, Mozambique, Peru and Sri Lanka;
see also Africa Watch: *Angola Civilians Devastated by Fifteen Year War* (5
February 1991), p 12; and Human Rights Watch/Africa, Children's Rights
Project: *Children of Sudan* (September 1995), pp 5, 56, 77–83
[58] Case study for El Salvador

Although many of these children and adolescents have entered voluntarily into participation in this type of activity, the whole concept of what is "voluntary" needs to be called into question. Various indirect coercive mechanisms have been used on these minors, such as physical protection, the stimulation of machismo, the symbology of concepts such as defense of the fatherland, the heroic nature of enlistment, revenge and adventure. Or through economic, cultural and social considerations such as belief in the justice of the cause, social pressure, provision of food etc. Therefore, the dividing line between voluntary and forced participation is very imprecise and ambiguous.[59]

The previous section considered the various methods by which recruits, particularly children, are selected against their will by armed forces or groups or their agents. This section considers why some children make a positive choice to join, or are encouraged or constrained to volunteer by force of circumstances, or where the choice is made on the child's behalf by the family. As much of the material illustrates, too great an implication of freedom of choice should not be associated with the term "voluntary" in this context. Nevertheless, certain motives and causes for voluntary recruitment can be identified and may be broadly grouped as cultural, in search of protection, ideological, and economic and social.

Cultural Reasons for Volunteering: In some situations, participation in military or warlike activities is glorified. Children are raised to revere military leaders of the past, and to look on military induction as a sign of manhood. At least to certain sections of opinion, the younger the age at which the child is involved in military activity, the more laudable. From the point of view of the child, the military life or the glamour and prestige of a military uniform may be attractions. Others may be persuaded to join by their parents in order to keep them off the streets, for political or patriotic reasons, or through a belief that military discipline is beneficial for the child.[60]

[59] Case study for Peru (translated from the Spanish)
[60] Case studies for Burundi, Burma/Myanmar, Colombia, El Salvador, Ethiopia, Guatemala, Liberia and Nicaragua

Khmer Rouge fighters in "The Children's Army" in Galaw, Cambodia.
Photo: Pica Pressphoto.

In some cases the young and especially the illiterates believed they gained prestige by belonging to an 'armed faction' with resulting power and authority which are unparalleled with what other children can achieve ... many become drunk with power and often abuse it.[61]

Children may succumb to the seductive lure of being able to command others, and of no longer being prey to their dictates ... To be a soldier is to occupy a position of great honour and self-sacrifice. The emotional pull of such prestige should not be underestimated.[62]

The charm of being part of an army, the weapons, the uniforms and the recognition of permanent role models were some of the reasons for enlisting ... The commandantes, for the children and young people, were the male figures which exercised the greatest attraction ...[63]

The need to distinguish themselves, to prove themselves as men, to become popular among their peers and neighbours was the motivation of the child soldiers mostly from urban settings; they would walk around the city with guns and in camouflage uniforms, perceiving themselves as important and older.[64]

Sometimes the incentive to become involved in the conflict comes from peer pressure rather than from the wider society. Some join simply because their friends have joined, particularly in urban areas and garrison towns. Children join armed opposition groups or participate in the conflict for the "adventure", "attracted by the sheer fun of belonging", or in order to become "famous and admired".[65]

Value systems which endorse bearing arms as a mark of masculinity can also draw, or push, youngsters into armed opposition groups, especially where linked to a tradition of "blood revenge", which requires even children

[61] Case study for Liberia
[62] Case study for Burma/Myanmar
[63] Case study for Colombia (translated from the Spanish)
[64] Case study for the former Yugoslavia
[65] Case study for Colombia, El Salvador, Ethiopia, the Intifada, Liberia, Sri Lanka and the former Yugoslavia; see also case study for Sierra Leone in M. McCallin: *The reintegration of young ex-combatants into civilian life* (ILO, Geneva, 1995)

60

to avenge the deaths of members of the extended family or even more distant kinsmen.[66]

Traditional values may also allow or encourage the use of existing youth groups, such as scouts, or the establishment of youth movements specifically to instil military values, commitment to the cause, and sometimes to provide familiarity with weapons, for example how to dismantle and clean them.[67]

In many cases, sections of the population, for example, those living in certain areas, or belonging to certain ethnic groups may feel a general, passive loyalty to the armed opposition group, or the wider movement, rather than to the government. This, quite distinct from any passionate involvement in the cause or tradition of blood revenge, may result in a sense of obligation simply to replace, rather than to revenge, a father, brother or other relative already lost in the struggle.[68]

In Search of Protection: Sometimes it is a desire for revenge that motivates the volunteer:[69]

> ... these youngsters may wish to avenge the genocide of their tribes, the torture and murder of their parents and the multiple rape (in the case of females) to which they were subjected.[70]

> Torture, abuse killing, deprivation and humiliation are all experiences which can create desires for revenge. Others include bombings, harassment, detention, and intimidation by security forces.[71]

However, it is remarkable how rarely in the case studies the children themselves cite revenge as a motive for volunteering.

---

[66] Case studies for Chechnya and Ethiopia
[67] Case studies for Afghanistan, Lebanon and Sri Lanka
[68] Case study for El Salvador
[69] Case studies for Burma/Myanmar, the Intifada and Liberia
[70] Case study for Liberia
[71] Case study for Sri Lanka

We came across only one soldier who explicitly gave the desire for revenge as his primary motivation.[72]

Blood revenge, like military tradition, seems to be a reason why families and communities "volunteer" their children. For the children themselves, far more often than revenge the motive seems to be their sense of vulnerability: enrolment in the government armed forces may be seen as a means to protect themselves and their families from harassment, whether by armed opposition groups or by the government forces themselves.[73]

What emerges as the single major reason why children volunteer for armed opposition groups is their own personal experience of harassment by government armed forces, including torture, loss of home or family members, enforced exile within the country or out of the country:

... most of the young individuals join [the opposition group] owing main-ly to [a] reaction to extensive human rights violations which they have personally witnessed from [the] security forces and most specifically, [the] systematic scorched earth policy ...[74]

Many of the youngest to join [the opposition] ranks were children, who had lost both parents in the war, either through death or exile, and had no-one else to take care of them. Many had seen their parents captured and/or tortured, even assassinated by army soldiers, their houses burnt, properties destroyed or robbed. They joined to look for protection.[75]

Young [ethnic] males often joined the combatants in order to avoid arbitrary detention. [Government] security forces in [the territory] often detained young males between the ages of 14 and 18 as potential combatants, in order to prevent them from joining the rebel forces.[76]

---

[72] Case study for Cambodia
[73] Case studies for Burma/Myanmar, Chechnya, El Salvador, the Intifada, Liberia, South Africa, Turkey and Uganda; see also case study for Sierra Leone in M. McCallin: *The reintegration of young ex-combatants into civilian life* (ILO, Geneva, 1995)
[74] Case study for Turkey
[75] Case study for El Salvador
[76] Case study for Chechnya

This would also appear to be the case where the involvement of the children in the conflict is spontaneous rather than as a result of recruitment:

> *They could not avoid the negative psychological consequences of harassment, humiliations, injuries, detentions, torture, or the death and injury of friends and relatives, the sealing or destruction of their homes, the nightly raids into homes in the refugee camps, seizing sleeping suspects and beating family members, particularly fathers – normally at the top of the family hierarchy – in full view of everyone else.*[77]

Others join because they feel it is better to die fighting than defenceless:[78]

> *It could become a much more attractive option to being helpless and afraid at home.*[79]

Little evidence was presented to suggest the converse: that mistreatment of civilians or recruits by armed opposition groups increases the number of volunteers entering government armed forces, merely that this reduces the number of volunteers into the group itself. This may be the result of an effective conscription system, making volunteering for government forces unnecessary, or may in part reflect the fact that armed opposition groups usually operate in limited areas and sporadically and therefore it is easier to avoid their activities by temporary or permanent relocation. On the other hand, the only legitimacy which can be claimed by such groups comes from popular support, and they are often aware of the need to keep on at least reasonable terms with the local population and sometimes make considerable efforts to cultivate them. Clearly, where an armed opposition group treats the civilian population and its own recruits well, it increases the number of volunteers entering its own ranks.[80]

---

[77] Case study for the Intifada
[78] Case study for El Salvador
[79] Case study for Sri Lanka
[80] Case studies for El Salvador and the Philippines

Ideological Reasons for Volunteering:

Some feel that children are attracted by a stark black and white vision of the world offered by the [armed opposition group]. However, there is information that some young recruits ran away from home to join the movement for the pleasure of riding a motorbike or tractor.[81]

Some children volunteer for armed opposition groups because they believe in what they are fighting for. This may be synonymous with the stated aims of the group: the holy war, the fight for freedom, freedom to practise their religion, the right to occupy their ancestral lands, ethnic liberation, political liberty. It may, on the other hand, be a general desire for social justice and opposition to poverty, corruption and militarisation, which need not imply a wholehearted endorsement of the ideology of the group's leaders, or stem from a sense of alienation – what other purpose is there in living? The children's commitment to the opposition cause may have been instilled in them throughout their upbringing and be reinforced by an idealisation of the culture of violence.[82]

Those who die for the cause may be considered as martyrs. Children and young people who participate are deemed to be "martyrs" or "honoured fighters", and the community values their participation and is proud of their achievements (even if their own families may be ambivalent, worrying for their safety). The cult of martyrdom plays a particular role in relation to recruitment and training of adolescents – particularly girls – for suicide attacks.[83]

In only one case study[84] is any evidence given of such loyalty or ideologically-based commitment to government forces, but it must be remembered that most of the case studies cover internal conflicts which to a greater or lesser extent pit governments against at least some of their own people. It is reasonable to assume that the situation would be different in the case of international conflicts.

[81] Case study for Sri Lanka
[82] Case studies for Afghanistan, Burma/Myanmar, Chechnya, Colombia, Ethiopia, Honduras, the Intifada, Lebanon, the Philippines, South Africa and Sri Lanka
[83] Case studies for Lebanon and Sri Lanka
[84] Case study for the former Yugoslavia

In some instances the armed opposition group has the support of the community and there is a high level of community involvement directly or indirectly: cultivating and distributing food for the combatants, caring for the wounded, informing about the movements of the army, hiding guerillas and so on. In these cases the communities generally approve of the participation of children and even support it. In any case, since in poor rural families, children, sometimes from as young as four, are actively involved in domestic and agricultural work, they are unlikely to remain outside a conflict which the community supports. Similarly, where 16 year olds are expected to work, and are treated as adults, the community also expects them to be active participants in armed conflicts.[85] The whole family may in fact be participating in the armed opposition group and be proud of their children's participation in the "holy war defending their native land and ideology".[86]

The armed opposition group may put its case by way of speeches calling for volunteers or use other methods of persuasion or indoctrination:[87]

... the guerillas arrived, killed a cow for the meal, there was a fiesta with the peasants – coffee-pickers – they stayed for several days and talked with the people.[88]

[an opposition] recruitment unit ... would enter an unguarded hamlet or village, gather all the villagers in a convenient area, propagate on behalf of their organisation and [its aims] and then ask for the local youth to join them.[89]

Sometimes recruitment has an even stronger ideological element:

The procedure for recruitment begins with the [opposition] regular eyeing individuals in the communities in the rural areas, who have made

---

[85] Case studies for Afghanistan, Chechnya, Colombia and El Salvador

[86] Case study for Afghanistan

[87] Case studies for Colombia, Guatemala and Turkey; see also Human Rights Watch/Africa, Children's Rights Project: *Children of Sudan* (September 1995), p 85

[88] Case study for Colombia (translated from the Spanish)

[89] Case study for Turkey

an impression on them. Usually, these would be individuals who are open to their cause. The [opposition] regular then approaches the potential recruit and starts an ongoing discussion on the [opposition] – its objectives, vision, rules. Once s/he agrees to join the [opposition] s/he undergoes a six months to one year trial period.[90]

Economic and Social Reasons for Volunteering:

For the refugees, internally displaced, homeless, orphaned and fearful, joining an armed group would appear an attractive option. Inadequacies in education and lack of hope in the future could make the [armed opposition group] an attractive proposition to a youngster. The [armed opposition group] could appear a disagreeable but known entity as opposed to an unknown, feared alternative of displacement and homelessness. Sometimes parents may also encourage children to join out of desperation when there is hunger and poverty in the family.[91]

In a very large number of cases the motivation for volunteering may be to find some means of survival or support, sometimes at the most basic level of feeding the family.[92] The alternative to enlistment may be unemployment.[93] Where it is traditional to have weapons at home and children are familiar with their use, the military option may be the preferred route for solving economic and other problems. This may lead job-seekers, particularly in areas of high unemployment where the army is the major employer, to conceal their real age in order to be recruited.[94]

However, it may be the family, rather than the individual child, who bows to economic pressure, especially where the army pays a percentage of

[90] Case study for the Philippines
[91] Case study for Sri Lanka
[92] Case studies for Afghanistan and Cambodia; see also case study for Sierra Leone in M. McCallin: *The reintegration of young ex-combatants into civilian life* (ILO, Geneva, 1995)
[93] Case studies for Afghanistan, Burundi, Burma/Myanmar, Cambodia, Chechnya, Colombia, El Salvador, Ethiopia, Lebanon, Liberia, Nicaragua, the Philippines, South Africa and Sri Lanka
[94] Case studies for Chechnya and Lebanon

a minor soldier's wages directly to the family. Similarly, armed opposition groups sometimes offer wages or personal allowances or various forms of subsidy for family members, including provision of food, transport or medicines; in one case there is also the possibility of access to work permits in an adjoining country.[95] Sometimes the economic inducement is not of pay, but vaguer promises of riches, or the knowledge that

> *the guns in their hands allow them to extract food forcibly from the civilian population.*[96]
>
> *In situations of deprivation, the gun can also become an entry point to food and survival.*[97]

In some cases, the economic motivation is for more than pure survival. The army has traditionally been one of the few routes of upward mobility in some societies (one of the few positions in society where power can be exercised over others through a post that is neither inherited nor attained through merit), and those poor but better educated children who therefore have prospects of becoming officers and earning a much higher salary may stay beyond the minimum period of service. Involvement in armed opposition groups may hold out to volunteers the prospect of becoming "a commander",[98] or may be a reaction to the frustrations of "educated, unemployed youth".[99]

Children who volunteer for economic reasons are particularly prone to deception. The benefits which they anticipate may not be forthcoming, with inadequate food and clothing.[100] Some specific examples are of children who were led to believe that by joining they would learn a trade and thus be able to earn their living, or were offered paid work with food and lodging clearing barricades from the streets of the capital but were sent to combat units.[101]

[95] Case studies for Cambodia, Guatemala, Lebanon, the Philippines and Sri Lanka
[96] Case study for Liberia
[97] Case study for Sri Lanka
[98] Case study for Burma/Myanmar, Colombia, El Salvador and Turkey
[99] Case study for Sri Lanka
[100] Case study for Burma/Myanmar
[101] Case study for Nicaragua

Sometimes volunteering provides a refuge which is not basically economic. One case study reports an interview with a boy who joined government armed forces in order to escape criminal charges[102]; another that
young criminals join "to have strong support of armed groups for robbery,
looting and other misconducts."[103] Girls may volunteer for armed opposition groups in order to escape from a marriage or an imminent marriage,[104] or conversely be encouraged by their parents to join because of
poor marriage prospects.[105]

[102] Case study for Burma/Myanmar
[103] Case study for Afghanistan
[104] Case study for Ethiopia
[105] Case study for Sri Lanka

CHAPTER 3

Who are the Child Soldiers?

Gathering and presenting facts on the topic of child combatants is constrained by the fact that reliable sources of information do not wish to be quoted for fear of ... reprisal. In addition, all reports, however carefully formulated, tend to be dismissed as propaganda by the group criticised.[106]

Where do the child soldiers come from? Irrespective of the method of recruitment, the overwhelming impression from the case studies is that the answer is the same. Firstly, from the poor or otherwise disadvantaged sections of society. Secondly, from the actual conflict zones themselves and thirdly, over and above the overlap with the other categories, from those with disrupted or non-existent family backgrounds.

THE POOR AND DISADVANTAGED

As is shown in Chapter 2, in practice, conscription frequently shades into forced recruitment. Even where this is not the case, however, the incidence is often uneven. The better-off families may send their children abroad, at least ostensibly, for education. While out of the country they will not be liable for military service and they may take pains not to return until the danger is over. Indeed, sometimes the whole family may relocate abroad.[107] The resulting loss to the society of some of its wealthier and better-educated elements may be permanent.

---

[106] Case study for Sri Lanka
[107] Case studies for Afghanistan, Lebanon and South Africa

Within the country, corruption may be more or less institutionalised as a means of avoiding conscription:

> *It must be made clear that the compulsory nature of military service does not, in practice, apply to adolescents from the middle and upper sectors of society who, by means of money or influence, manage to have themselves declared "unfit" or simply "not selected" in the selection procedures for recruits.*[108]

Or the means of evasion may be indirect: one case study[109] records that it is possible to purchase a military service record, the essential documentary proof that the required service has been undertaken. By contrast, members of various groups who are legally exempt (for example, the only sons of widows or separated mothers, heads of households, or members of indigenous groups) often have difficulty in producing documentary evidence to establish their status. Where conscription is organised on a quota basis, the scope for "buying out" is obviously greater. One case study[110] reported an extreme institutionalisation of this, which was termed the "conscription lottery". Under this system, the names of those eligible for conscription were drawn by lot but, for a fixed price, the family of the person whose name was drawn could pay the army to draw again.

As far as forced recruitment is concerned, even if within recruiting raids individuals are seized at random, the direction of the raids themselves is far from random. Overwhelmingly they target gatherings of the poorer, more disadvantaged groups in society. Nor is this just because such groups tend to cluster together in places where they can easily be rounded up. Government recruiters typically target those who are perceived as a threat: directly (potential recruits to armed opposition groups) or indirectly (supporters or potential supporters of opposition groups). These may be specific ethnic, racial, indigenous or religious groups; they may (as discussed below) be the inhabitants of particular areas; or they may simply be the poorest and most disadvantaged sections of society. Recruiters naturally

---

[108] Case study for Peru (translated from the Spanish)
[109] Case study for Colombia
[110] Case study for Burma/Myanmar

prefer to concentrate on those who can mount the least effective resistance or challenge, which means the most disenfranchised groups. Over and above these considerations, recruitment may form one weapon in a general campaign of intimidation, repression, or general disempowerment and disruption. There is a consistent pattern that the more influential classes in society and particularly the wealthier parts of urban areas are immune from such recruitment.[111]

Once seized by a "press gang", the recruit has little chance of escape. In some cases there is enough respect for age limits that those who can produce documentary proof of age may be released. However, the typical victim of forced recruitment is unlikely to be able to do so. Often the production of proof depends on the family managing to trace the recruit. However, recruits are typically posted as far as possible from their home area[112] which not only breaks the link with the local community and makes it harder to desert but also makes it more difficult for families to trace the recruits, and may make the search prohibitively expensive. In addition to the physical and financial difficulties entailed, the anguish of the family searching for the child who has disappeared in this way should not be overlooked.

It may be possible to evade recruitment. One case study reports:

As people riding in cars were usually safe [from press ganging], employers, friends and relatives with cars would pick up men and boys from school, university or place of work when they heard that there was a danger of "afesa" [press gang].[113]

Whether or not this is intentional, the fact that those in cars are safe once more helps to shield the relatively prosperous.

There is sometimes the possibility of challenging under-age recruitment through publicity or legal channels; some of the rare instances when chal-

---

[111] Case studies for Burma/Myanmar, Cambodia, El Salvador, Ethiopia, Guatemala, Mozambique, Nicaragua, Paraguay and Peru

[112] Case studies for Afghanistan, Bhutan, Burma/Myanmar, Colombia, Guatemala, Honduras and Nicaragua

[113] Case study for Ethiopia

lenges have been successfully mounted are discussed in Chapter 8. However, to mount such challenges may require a degree of organisation, education and money.

Finally, and more consistently, there is the possibility of buying oneself out:

> *The main strategy for keeping children out of the army was to bribe the recruiting units or local commanders or village chiefs, depending on who was in charge of the recruitment or conscription. Paying the army not to take a child or adult was common practice as was the practice of paying a poor family or individual to take one's place.*[114]

> *The sons of the well-off families did not run the risk of being recruited even in peacetime, and less at the time of the war. If a son of an upper- or middle-class family was taken by mistake in ... the forced recruit-ment, he was quickly released when the parents showed up to demand him or to pay a bribe for his release.*[115]

> *Our parents had no idea what happened to us. They weren't told any-thing and neither were we. Some people had money to pay off the ... offi-cers, but most couldn't.*[116]

Sometimes buying out may take a disguised form. One case study[117] records that when families do manage to trace illegally recruited children they may have to pay for documents to prove the child's age and be charged for the food, lodging and clothes of the under-age recruit while he was in the armed forces.

All these factors operate in the same direction. Children from the more prosperous, privileged classes suffer far less risk of recruitment into govern-ment armed forces, and if by chance they are picked up they are more likely to be released. Thus they rarely become child soldiers.

Nor is the focus on the disadvantaged unique to government recruiters:

---

[114] Case study for Cambodia
[115] Case study for El Salvador
[116] Case study for Burma/Myanmar
[117] Case study for Guatemala

It has been reported that most of the child combatants are school drop-outs and from poorer families. The elite and their children are accorded favoured treatment in return for complicity.[118]

As the fighting among armed forces became intensified and the rebel forces sustained losses in the process, their leaders imposed a quota for new recruits who in most cases comprised street and other disadvantaged children to augment their manpower.[119]

What of volunteers? It is precisely the poor and disadvantaged groups who are most at risk from forced recruitment who also have the strongest economic and social incentives to volunteer. The difference is that military service to the volunteer tends to be seen as a route out of destitution rather than a further aspect or consequence of it. Even more than poverty, educational deprivation is the hallmark of the child volunteer. Limited education or educational opportunities are frequently mentioned in the case studies as a reason for volunteering.[120] Volunteers may be those who

despair of education and see the armed forces as an adventuresome opportunity to escape from the boredom and tedious routine of formal education.[121]

In other circumstances they may have been attracted by promises of education, study grants, or overseas travel.[122] Lack of education is even more characteristic of volunteers than of forced recruits.

The outcome of all this can be summed up by quoting one analysis of

---

[118] Case study for Sri Lanka

[119] Case study for Liberia

[120] Case studies for Afghanistan, Cambodia, Chechnya, Colombia, the Intifada, Lebanon, Mozambique, Nicaragua, the Philippines, South Africa and the former Yugoslavia; see also case study for Sierra Leone in M. McCallin: *The reintegration of young ex-combatants into civilian life* (ILO, Geneva, 1995)

[121] Case study for Ethiopia

[122] Case studies for Afghanistan, Chechnya, Colombia, the Intifada, Lebanon, Mozambique, Nicaragua, the Philippines, South Africa and the former Yugoslavia; see also Human Rights Watch/Africa, Children's Rights Project: *Children of Sudan* (September 1995), pp 77–83

the background of demobilised child soldiers, forced recruits and volunteers:

> *... the boys mainly came from poor peasant families in isolated rural areas or from the conflict zones. The poverty of their homes bordered on destitution to judge by the clothing, type of house, environmental conditions and educational level of the families. The child soldiers from urban areas came from homes where the head of the family was a woman; they were the sons of cooks, fruitsellers, small traders. These families had numerous children and they were obviously poor, judging by the materials of which the house was made, their clothing and the areas they lived in ...*[123]

A uniquely disadvantaged sub-group in society, and hence a particularly fruitful source of child soldiers, are the inhabitants of camps for refugees or internally displaced persons.[124] In particular, displaced groups are likely to regroup themselves according to their religion or ethnicity, forming ready recruiting grounds for armed opposition groups if these themselves are so based – the degree of recruitment being correlated with the perceived level of threat from other groups. Furthermore, their general problems are often compounded by inadequate economic provision.[125]

> *Returnees from the border camps and other landless or internally displaced persons are the most vulnerable people in [the country] and it is from these families that boys leave to join up.*[126]

THE INHABITANTS OF THE CONFLICT ZONES

Where a conflict has persisted over many years, the children of the conflict zone will inevitably be among the poorer and most deprived members of

[123] Case study for Nicaragua (translated from the Spanish)
[124] Case study for Cambodia; see also Human Rights Watch/Africa, Arms Project: *Angola: Arms Trade and Violations of the Laws of War Since the 1992 Elections*, (NY: Human Rights Watch, 1994) p 87, and Human Rights Watch/Africa, Children's Rights Project: *Children of Sudan* (September 1995), pp 4, 56, 60–61
[125] Case study for Lebanon
[126] Case study for Cambodia

74

society and, therefore, particularly prone to recruitment. The local economy is affected by the conflict; educational provision suffers and the wealthier part of the population may well have relocated or at least sent its children to a safer part of the country[127] thereby adding to the economic decline of the region. Also in the conflict zones an unusually high proportion of households will through death and injury have ended up with children as the main income earners.[128] Not only is this likely to be correlated with extreme destitution but, with the economy in tatters, child bread-winners are likely to look to the only growth industry – the armed forces or groups.

However, the population of the conflict zones are at risk over and above their poverty. From the point of view of government armed forces, they are particularly likely to be targeted for recruitment as actual or potential supporters of armed opposition groups. If the armed opposition group practices forced recruitment, it is the population of the conflict areas to whom they have access and who are therefore at risk. It can happen that the same village is obliged to supply a quota of recruits to both the government and the opposition.[129] Children from the same village can find themselves fighting on opposing sides:

> *(One child soldier) tells how he and the other adolescents of his village ... were in the thick of the action all along the demarcation line and in some cases crossed that [same] line to go to their village for the weekend, returning for the working week to their places on the demarcation line, opposite youngsters (from the same region) who they sometimes knew (on the other side).*[130]

In the conflict zones too, all the factors which predispose children towards volunteering are present in acute form, especially the lack of alternative opportunities and the tendency to view the conflict, and participation in it, as the normal way of life (also illustrated in the above quotation). This

---

[127] Case studies for Afghanistan, El Salvador, Lebanon, Liberia, South Africa and Sri Lanka
[128] Case study for Afghanistan
[129] Case study for Burma/Myanmar
[130] Case study for Lebanon (translated from the French)

Chechen boys stand guard at a Chechen check-point west of Grozny 1995.
Photo: Oleg Popov, Reuter/Pressens Bild.

includes situations where family members live in the army camp with the soldiers and the children therefore consider it natural for them to join up too.[131] Above all, it is in the conflict zones that children are likely to have the direct personal experience of abuses by government armed forces identified in Chapter 2 as the main reason for their volunteering for armed opposition groups. The victims of forced resettlements in the conflict zones are particularly prone to choose the alternative of seeking security and stability in membership of an armed opposition group.[132]

The combination of factors in the conflict zones makes the involvement in the conflict more widespread throughout the population of such zones, including children. However, it seems likely that they specifically tend to increase the recruitment of children and of steadily younger children, in particular because there is competition for the available manpower which is diminishing as the conflict continues. Perceiving the danger that children may be recruited into armed opposition groups, government forces attempt to pre-empt this, which is counter-productive, since it tends to increase volunteering into the armed opposition group.[133] Such volunteering is primarily occasioned by harassment or a sense of resignation to the inevitability of being forced to take a part in the conflict and therefore a desire to have some choice in the matter of sides and some belief in the cause for which they are fighting. However, it is also strongly influenced by the perception that treatment of and conditions for recruits in the armed opposition groups are better than in the government forces. At other times in the conflict zones, the armed forces or armed opposition groups may for humanitarian reasons pick up unaccompanied children who, however, end up joining the fighting ranks.[134]

The way in which the conflict becomes an integral part of the lives of some children is shown by the references to their changing sides, either as

---

[131] Case study for Cambodia
[132] Case studies for Burma/Myanmar and Ethiopia
[133] Case studies for Afghanistan, Burma/Myanmar and El Salvador
[134] Case study for Liberia; see also case study for Uganda in M. McCallin: *The reintegration of young ex-combatants into civilian life* (ILO, Geneva, 1995)

a result of defection or of capture in battle.[135] When this is a case of defection, the motive would usually come under the heading of "ideological"; at other times it may be simply that the children "have got accustomed to the gun culture and do not have skills for any other type of job"[136] or that they may feel the need for protection: "Some of these children may have committed atrocities ... and fear reprisals."[137]

SEPARATED CHILDREN

No group in society is as vulnerable as children separated from their families for whatever reason. Inevitably the conflict zones contain a particularly high proportion of children who have lost or are separated from their families. All the evidence in the case studies suggests that these groups, and other children whose family background is unstable or disrupted – whose fathers have been recruited, detained or killed in the conflict leaving the mother as the head of household, illegitimate children, children of divorced or separated parents and of single-parent families in general – are even more prone than their peers to become child soldiers.[138]

There are various reasons for this. One is that the family provides a measure of physical protection and assistance in strategies for avoiding recruitment. Children with no families have noone to send them away into safety. Accompaniment by adults is sometimes an effective protection against forced recruitment. By contrast, it has been reported that unaccompanied children even from solid family backgrounds, and even when running errands for the family, have been picked up and designated as street children.[139] (Although sometimes whole families are kidnapped to

---

[135] Case studies for Afghanistan, Burma/Myanmar, Cambodia, El Salvador, Liberia, Mozambique and Uganda

[136] Case study for Uganda

[137] Case study for Uganda

[138] Case studies for Burundi, Burma/Myanmar, Cambodia, Chechnya, Colombia, El Salvador, Ethiopia, the Intifada, Lebanon, Liberia, the Philippines, Turkey and the former Yugoslavia; case study for Uganda in M. McCallin: *The reintegration of young ex-combatants into civilian life* (ILO, Geneva, 1995)

[139] Case studies for Burma/Myanmar and Ethiopia; see also Human Rights Watch/Africa, Children's Rights Project: *Children of Sudan* (September 1995), p 56

the armed opposition base, where the children are given military train-ing.)[140]

Unaccompanied children are an attractive target for a recruiting party because on the simple physical level they will be unable to resist as effec-tively as adults. Once forcibly recruited, such children are unlikely to have anyone to make enquiries, let alone to mount any form of challenge or to pay a bribe. Then again, the family sometimes – though obviously not always – acts as a repository of alternative values:

> *Educated families/parents saw the military struggle as barbaric and uncivilised and therefore did everything to ensure the non-involvement of their children. Some families/parents because of their religious back-ground ... did not allow their children to participate in it. While some parents, because of their experiences [from] the devastating effect of war, stopped their children from participating in the conflict.*[141]

Without families as sources of discouragement, children are more at the mercy of militarist cultures, peer pressure and the inability to conceive of life outside the confines of the conflict. They are also more vulnerable to trickery such as false accusations of theft, leaving the choice of army or prison.[142] Obviously, family pressure is likely to weigh more heavily with the child, the more stable and cohesive the family background itself is.

Finally, it is in the search for a substitute or replacement family that children in these categories are particularly liable to join or attach them-selves to armed forces or armed opposition groups:

> *For those without a family, the militia acts as a large family with a leader and brothers in arms.*
>
> *The testimony of former child soldiers in prison reveals why they became involved in the militia: [boy] 16, is illiterate and ran away from home because of on-going differences with his father. [Boy] 15, could not go to school as his father could not afford it (having another child handi-*

[140] Case studies for Mozambique and Peru
[141] Case study for Liberia
[142] Case study for Cambodia

capped by the war) and encouraged him to go and find work, which led him to leave home and join the local militia. [Boy] 16, the child of his father's first marriage, constantly in conflict with his stepmother, left home to work in a bakery where his workmates, who were in the militia, encouraged him to join.[143]

"... when I was going [to the army base] there were two minors, one of 15 and one of 11 or 12. The 11 year old went to [the army base] because he had no family ..."[144]

... a 12 year old who was conscripted at ten years of age with his older brothers. This was done at the behest of his father who had remarried after the death of his first wife and no longer wanted the responsibility of raising his sons.[145]

GENDER

– families will hardly allow their girls to marry former child soldiers...[146]

Female guerillas can be seen these days in almost all [opposition] camps and gradually they are taking on more active duties than only cooking and/or cleaning.[147]

The girl soldiers were stationed at the front in all military actions, and so bore the brunt of any casualties.[148]

Nowhere is the influence of cultural expectations seen more strongly than in relation to the gender of child soldiers. Reports tend to mention "children" and very careful reading is sometimes necessary to ascertain whether the children are of both sexes or only boys.

The main impression given by the case studies is of a contrast between

---

[143] Case study for Lebanon (translated from the French)
[144] Testimony of a minor recruited but then freed by the army, quoted in the case study for Peru (translated from the Spanish)
[145] Case study for Cambodia
[146] Case study for Afghanistan
[147] Case study for Turkey
[148] Case study for Cambodia

government armed forces and armed opposition groups, with the former overwhelmingly if not exclusively male, while the latter tend to involve both males and females of all ages. There are exceptions in both directions: in Peru both men and women are covered by the Law on Compulsory Military Service; in Afghanistan (see quote above) the unstated implication is that all child soldiers on all sides are boys.

In fact, the contrast between government forces and armed opposition groups is not as absolute as it appears at first sight. Whereas everyone living in a guerilla camp, except perhaps the smallest of the children, is self-identified as a member of the group, the equivalent is not necessarily true of the civilian population on which government armed forces depend. Behind the combatant forces there is usually a vast hidden army of support and ancillary workers who may or may not show up in official figures. These are often women and girls, and sometimes small children:

Women ... essentially provided support services such as secretarial and musical services. More and more women, however, were involved in these services as the size of the army expanded ... As the services were far removed from combat duty and were mainly carried out in urban areas an increasing number of young girls were looking at the armed forces ... as a means of winning their living in the context of widespread unemployment.[149]

One interviewed ex-military said: "Every officer used to have his assistant or servant, who was a cipote [little boy]." ... In general, women and female children ... performed "domestic functions" like cooking, washing, cleaning, doing secretarial work in administration etc.[150]

In Cambodia, large parts of the government armed forces function in a way which would normally seem more characteristic of armed opposition groups:

... families live in temporary huts constructed near the army base where the soldier is serving ... often the families will live in the camp itself with the soldiers ... and cook for them and eat with them ... Families are also

[149] Case study for Ethiopia
[150] Case study for El Salvador

to be seen in the military and provincial hospitals caring for the wounded and sick soldiers ... Women and girls have been used in support roles ... and [one] study gives an example of a soldier wife being used as a porter in a dangerous area. Women also fight on the government side, but these women are invariably married to soldiers or have family in the force and there has been no evidence of women under 18 being recruited.[151]

This said, the fact remains that women – and girls – are more to the fore in armed opposition groups, and are more likely to be combatants:

Unlike their counterparts in the [government] army, women in the [opposition] force usually assumed combat duties. In fact, some are said to be among the best fighters. Not an insignificant number of them were posted in commanding positions within the force ... again, not an insignificant number were believed to have been recruited while they were under age.[152]

The braver girls participated fully in war including a few who were commanders.[153]

... the commander of one of the self-defence units was a girl aged 16.[154]

Female children were not treated differently either, unless they requested to be assigned to a medical or kitchen unit, in which case the request was accommodated.[155]

For example, in El Salvador, where estimated government forces of 60,000 included only one short-lived battalion of 160 armed women and girls, of 2,000 demobilised opposition troops who were aged under 18, some 700 were girls. In Ethiopia it was estimated that women and girls formed between 25 and 30 per cent of the opposition strength. In Cambodia, on the opposition side, there is a report that "a group of 300 to 500 girls under

[151] Case study for Cambodia
[152] Case study for Ethiopia
[153] Case study for Uganda
[154] Case study for South Africa
[155] Case study for Chechnya

83

15 were kept together and given military training". Of 183 child soldiers recaptured from the armed opposition group in Uganda, 55 were girls. Of the ten interviews with members of the NPA opposition group in the Philippines, six were former child soldiers, and two of these were women. The involvement of women is extending to societies, such as the Kurdish population in Turkey or the Tamils in Sri Lanka, where women have traditionally played a subordinate role.

> *Many, many women come to the guerillas wanting to fight, a stream of them, but they in particular have been held back in their development by feudal society.*[156]

In Turkey there are reports that the PKK has organised a "children's battalion", comprising three divisions, two of boys and one of girls. This seems to reinforce the pattern discernable in the other cases quoted, namely that when girls are fully involved, they provide approximately one in three combat troops.

Although, in general, girl soldiers perform the same functions as boys, sometimes distinctions are made:

> *Soon after recruitment the girls are divided up and allocated to rebel men to be their "wives".*[157]

> *... the [armed opposition group] also recruits young girls into roles supporting the adult recruits, in preparation of food, attending to the wounded, washing clothes, without neglecting the relevant military training.*[158]

> *Girls are also enrolled into the opposition armed group and essentially take care of the kitchen work and the rendering of sexual services, except those who ... are broken into guerilla techniques.*[159]

> *... in the [opposition] camps, even the boys, already traumatised them-*

[156] Case study for Turkey
[157] Case study for Uganda
[158] Case study for Guatemala (translated from the Spanish)
[159] Case study for Burundi (translated from the French)

selves, violently sexually assaulted the girls, even threatening them with death or removal of food rations if they resisted.[160]

... girls took on the traditional household chores (looking after the youngest ones, cooking ...), and the older ones were 'promoted' to marriage.[161]

It is not clear whether marriages were forced or merely encouraged. Such marriages for young girls may complicate demobilisation, when they may be torn between staying with their husbands or returning to their families.[162] Others may be teased by their fellows if they return to school.[163] Involvement in sexual activity for girls (and indeed boys), whether dignified as marriage or not, entails for many the additional hazards of sexually transmitted diseases, HIV/AIDS, and pregnancy. The recruitment of young girls may be a deliberate attempt to provide "wives" free from HIV infection, thus "the criteria used for 'marrying' girls to rebel men seems to be signs of puberty."[164] More than one armed opposition group requires abortions if female members become pregnant.[165]

In Sri Lanka, there has been a tendency for girls to be used as suicide bombers. This is also the case in Lebanon where, otherwise, no evidence of female involvement in the conflict was given by the case writer.

---

[160] Case study for Mozambique (translated from the Portuguese)
[161] Case study for Mozambique (translated from the Portuguese)
[162] Case study for Mozambique
[163] Case study for Uganda
[164] Case study for Uganda
[165] Case studies for Honduras and Peru

CHAPTER 4

Treatment as Soldiers

As children, they are suspects and probable victims. As soldiers, they are too often killed...[166]

Generally, child soldiers in government armed forces receive the same treatment and training as adult recruits:[167]

All my sources agreed that child soldiers received no special treatment in the armed units. There were no differences in the treatment of adult and child soldiers.[168]

According to the testimony of the minors, they were treated like any other soldier; in the military units in which they served no one took account of the fact that they were under age.[169]

Child soldiers were not accorded any preferential treatment to that of adult soldiers. Nor were there any variations in the type of activity they performed on the basis of their age or method of recruitment.[170]

In a great number of the interviews conducted ... teenagers reported that their duties were similar to those of all rank and file soldiers. [Government] soldiers in particular stated that distribution of duties

---

[166] Case study for Turkey
[167] Case studies for Afghanistan, Burma/Myanmar, Cambodia, El Salvador, Ethiopia, Guatemala, Nicaragua and the former Yugoslavia
[168] Case study for the former Yugoslavia
[169] Case study for Nicaragua (translated from the Spanish)
[170] Case study for Ethiopia

had little to do with age. All ... we interviewed had been in active combat and had spent most of their tenure in the front lines doing the same fighting as the older soldiers. They had all fired weapons, and most had killed people.[171]

In those situations where the recruitment of children is illegal or unacknowledged it would be remarkable if this were not the case. In some respects equal treatment may be beneficial for the children in that they get paid and if their recruitment was in fact conscription, even though underage, they may be able to leave after the prescribed term of service, and receive the other benefits of normal service. (In armed opposition groups it is very rare for combatants, even children, to be allowed to leave unless invalided out).

However, the treatment of governmental recruits of all ages is often inhuman and degrading. This starts with initiation rites for new recruits ("baptism", "hazing") involving beatings, humiliation, being treated as servants, and being introduced to prostitutes and alcohol. The degree of violence and degradation involved not infrequently leads to death (including suicide), disablement or permanent physical, mental and emotional damage.[172] That such treatment is meted out in a state institution raises questions about human rights and governmental responsibility, as well as how such treatment affects the subsequent behaviour of soldiers. These are subjects which merit a study in their own right.

A few examples of such treatment suffice to illustrate the point:

... a succession of exercises lasting several hours to break the physical resistance of the recruit by provoking intolerable muscular suffering ... Hitting and body bruising with instruments such as machetes, sticks, rifle or pistol butts and yataganes (special clubs that leave no marks) on sensitive parts of the body; burning with cigarettes on various parts of the body or swallowing burning cigarettes; fist punches in the face, the nape of the neck, the ears; kicking in the legs and stomach and stamp-

[171] Case study for Burma/Myanmar
[172] Case studies for Burma/Myanmar, El Salvador, Guatemala, Honduras and Paraguay

ing on hands. Psychological pressures including threats, mockery and continuous insults ...[173]

... they are trained to become accustomed to sadism and, in order to lose their fear, they practise by cutting the throats of domestic animals and it is known that some of them are obliged to drink their blood. The internal violence which the recruits most fear is the verbal pressure exercised by the officers, pressure aimed at minimising their self-esteem and breaking their characters.[174]

Among the toughest exercises is "the ram" which consists of rolling on any surface which is stony, thorny, or asphalted etc. Your body ends up skinned because they have you just in your underpants. Also, while you are rolling, the man in charge of the squad kicks you or hits you with a club. Another difficult exercise is to be shut up in a waterproof tent, it consists of shutting up 20 or 40 recruits in the tent and then they throw in tear gas grenades and outside the corporals stand guard to make sure no one comes out until they give the order.[175]

Although such brutal treatment may be the normal experience of recruits, it may fall unequally on the child recruits who are less well-equipped physically, mentally and emotionally to withstand it:

... most child soldiers interviewed for this survey seemed to either fear or revere their commanders, and stated that they would follow their every order. Fear is a major determining factor in their obedience and performance. However, in extreme cases, when child soldiers could no longer tolerate their own, or others' mistreatment by their senior officers, they were driven to either suicide or murder.[176]

Or the child soldiers may be deliberately picked on:

[173] Case study for Paraguay (translated from the Spanish)
[174] Case study for Colombia (translated from the Spanish)
[175] Case study for Honduras (translated from the Spanish)
[176] Case study for Burma/Myanmar

"In most cases child soldiers received harsher treatment and were also blamed for faults of adult soldiers."[177]

"Irresponsible children" may be blamed by adult soldiers for acts of pillage carried out by the adult soldiers themselves.[178] The assignment of poor and inadequate food and medical care (as the lowest rungs on the military ladder) may be more serious for children, whose bodies are still growing and may be weakened by the exactions of military life:[179]

... the food is vile and also before and after eating we had to do knees-bends so that we would vomit our food ...[180]

The same recruit can do three shifts of guard duty in a single day, which means he can be standing up for nine hours, and exposed during the same period to guerilla attack.[181]

Treatment in armed opposition groups is equally inhuman in many instances, with those who cannot keep up or who try to escape being routinely killed so as not to reveal any secrets or to discourage others.[182] Even if such treatment is the same as for adults, "it is mostly children who suffer this fate as very few of them manage to stand up to the intensive training to which they are subjected, the military discipline, the diet and the unhealthiness of the surroundings."[183] Similarly, being less adept at looking after themselves, or standing up for their rights, they are more prone to die from starvation and preventable diseases contracted in the unhygienic conditions in which they live.[184]

However, this is not always the case:

Even the most ardent critics of the insurgent [opposition] forces could not find fault with the manner in which child soldiers were treated.[185]

[177] Case study for Afghanistan
[178] Case study for Lebanon
[179] Case study for Burma/Myanmar
[180] Case study for Honduras (translated from the Spanish)
[181] Case study for Colombia (translated from the Spanish)
[182] Case studies for Burundi, Peru and Uganda
[183] Case study for Burundi (translated from the French)
[184] Case study for Ethiopia
[185] Case study for Chechnya

Where children are well treated in armed opposition groups, this is because all members of the group are well treated, reflecting "the warmth of friendship/camaraderie";[186] the need to maintain the numbers and morale of the group, the apparent absence of forced recruitment, and the ethnic- and community-based nature of the conflict. Similarly, where whole families are involved in the armed opposition group, the children tend to be treated well since they are under the care of their own families.[187]

DIFFERENTIAL TREATMENT

Some opposition groups are exceptional in seeking to take account of the particular needs and abilities of children. This may be by annually evaluating the child's physical and mental state when deciding on tasks and promotions, according children a degree of emotional attention, with some being "adopted" by commanders, providing time for play, and some preferential food and accommodation. Some, at least when there is a truce or a lull in the fighting, provide basic education.[188] Education, however, may have an ideological content:

> ... a third group was made up of the "legions" or "pioneers of the future", children under ten years old, of both sexes, who lived separated from their parents and attended the "people's schools" where they were given courses in political indoctrination.[189]

Governments which legally recruit under-18s may also take special measures, for example, by having a policy of not sending them on active service[190] or "Recreational facilities are provided in an environment free from alcohol use. Their progress is carefully monitored by supervisory staff and there is ready access to chaplains, psychologists and other support staff."[191]

---

[186] Case study for the Philippines
[187] Case study for Afghanistan
[188] Case studies for Colombia, El Salvador, the Philippines and Turkey
[189] Case study for Peru (translated from the Spanish)
[190] For example, Australia and the Netherlands
[191] Response from Headquarters Australian Defence Force to enquiry by Quaker UN Office, Geneva

On the other hand, some armed opposition groups mete out especially cruel or dehumanising treatment to them. The use of children as executioners is not unusual,[192] although acts of ritual cannibalism appear to be the exclusive province of a small number of armed opposition groups seeking to imbue children with contempt for human life:

... the minors were obliged to cut the throats of those who had been declared culpable by the "people's courts". Often they were made to eat the entrails – heart, liver, kidneys – and to drink the blood of the rebels who were sentenced to death.[193]

During [the boy's] time with the guerillas he had the opportunity of witnessing many executions of boys and girls of 12 and 13. Later, with the passage of time, he carried them out himself and taught others to do it. The practice of drinking the blood of the dead was a great trauma for him; he used to dream that the dead person would come back to reclaim his blood.[194]

While, in the testimonies, there is mention of human targets to "toughen up" the children, it is important to add that in some cases these targets were parents of the child being trained. In other cases, the child could be forced to watch the ritual which might involve his/her parent. Such was the case of one child who stated: "In captivity, my father was used as a target during the final tests of boys who were being trained."[195]

Almost all the children at one time or other were ordered to torture, maim or kill other children or adults who attempted to escape or to destroy property like burning up houses. With intimidation, threats and sometimes torture, children were obeying like robots.[196]

[192] Case studies for Burma/Myanmar, Colombia, Honduras, Liberia, Mozambique, Peru and Uganda

[193] Case study for Peru (translated from the Spanish)

[194] Case study for Colombia (translated from the Spanish). Cannibalism being used as a rite of passage to combatant status by Renamo is reported in N. Boothby, P. Upton & A. Sultan, *Boy Soldiers of Mozambique*, Refugee Participation Network 12 (March 1992), pp 3–6, p 5

[195] Case study from Mozambique (translated from the Portuguese)

[196] Case study for Uganda

... reports indicate use of children as young as 10 years as assassins. Children of 12–14 years of age are also reported to have been used to massacre women and children in remote rural villages.[197]

There are numerous reports of children being given drugs and/or alcohol routinely or before battles.[198] Sexual abuse (particularly of girls) is not infrequently reported though by no means universal.[199] By contrast, some armed opposition groups ban the use of drugs and alcohol and demand respect for women.[200]

FUNCTIONS PERFORMED

In armed opposition groups children often start out in support rather than combat functions. They may, indeed, be "associated with the military" rather than an actual part of it, being alone and destitute and having attached themselves to individual soldiers or troops, and becoming servants or guards in exchange for food and shelter.[201] In many situations children are used for guard duty, patrolling and manning check-points:[202]

... most children ... were used normally for behind-the-front activities before being given military training and accepted as full-time combatants.[203]

As soon as they were recruited they would be given the duty of carrying luggage, for example looted foodstuffs including chickens, ammunition, property as well as injured rebels ... Most times the children did guard duties or worked in the gardens or were sent to hunt for wild fruits and

[197] Case study for Sri Lanka
[198] Case studies for Afghanistan, Burma/Myanmar, El Salvador, Honduras, Liberia, Mozambique and Peru; see also Amnesty International, "Sierra Leone: Human rights abuses in a war against civilians" (13 September 1995), p. 25
[199] Case studies for Afghanistan and Mozambique; case study for Uganda in M. McCallin: *The reintegration of young ex-combatants into civilian life* (ILO, Geneva, 1995)
[200] Case studies for El Salvador, the Philippines and Uganda
[201] Case study for Rwanda
[202] Case studies for Bhutan, Burma/Myanmar, Cambodia, El Salvador, Ethiopia, Guatemala, Liberia, Mozambique and Peru
[203] Case study for Turkey

*vegetables or looted food from gardens and granaries. But if attacked
by the enemy, they were expected to fight back with the adult rebels.
When there were fierce battles and shortages of fighters, child soldiers
would be taken to the war front to fight alongside the adults.*[204]

In any case, they were being prepared to play an active part when older:

*Almost all the children who stayed for over one month [before escape or
capture] were given military training of parade, taking cover, assem-
bling and dismantling a gun, shooting, and offensive tactics.*[205]

*Children from 10 years of age could become combatants, although
mainly children from 14 or 15 years up participated in fighting with
arms.*[206]

*... around 1,000 children aged between 7 and 14 held a meeting during
which a decision was jointly taken to begin military organising activities
and establish childrens' committees which will be similar to the guerilla
organisation. The decision ruled that these committees will organise for
the time being on the training level and that, after the age of 14, volunteers would transfer to the ranks of the guerillas.*[207]

*Children have also been observed to spend one or two hours a day out
of school digging bunkers as a form of militarised civic duty and then
motivated to enlist and join the armed opposition group.*[208]

Numerous children are used as porters during conflicts. Often they carry
very heavy loads (food, up to 60 kg of ammunition, injured rebels) and
receive severe treatment at the hands of their superiors. In some cases, they
are savagely beaten and, if they become too weak to carry their load, liable
to be shot.[209]

[204] Case study for Uganda
[205] Case study for Uganda
[206] Case study for El Salvador
[207] Case study for Turkey
[208] Case study for Sri Lanka
[209] Case studies for Burma/Myanmar, El Salvador, Mozambique, Sri Lanka, Turkey
and Uganda

Many of the case studies refer to a special preference for using children as look outs, messengers and for intelligence work:[210]

There is also evidence that very young soldiers or civilians are used as spies in the disputed areas. They are generally given cows to herd and asked to wander around and to report back on any [opposition] activity they had witnessed. In [one area], for example, children herding cows were asked by the [opposition] to place home-made landmines (crude bombs made from fertilizer) on rural roads late at night or early in the morning. In some cases, the local government forces will use young civilians or very young soldiers to do the same near [opposition] villages. Very young soldiers are not noticed so much and can easily pass themselves off as normal peasant children and are used by both sides as intelligence gatherers. ... Children were used a lot to lay and to clear mines but we did not hear of them being used exclusively for this purpose.[211]

... they are likely to be less noticeable than adults. Because of their size, they can hide more easily, and they have more innocent appearances ... [An NGO] interviewed a 14-year-old [opposition] recruit in 1990, who told an autobiographical story that seemed normal. The next day he was caught in the ammunition supply room ... it was discovered that the [government forces] recruited him as a spy and were threatening him that they would kill his grandmother who had taken care of him since he was a baby. He was later executed by the opposition group concerned for spying, despite his young age.[212]

[In the opposition forces] younger children were considered more suitable as messengers, since they would not raise suspicion ... In the

[210] Case studies for Afghanistan, Burma/Myanmar, Burundi, Cambodia, Colombia, El Salvador, Ethiopia, Guatemala, Liberia, Mozambique, Nicaragua, Northern Ireland, Peru, the Philippines, South Africa, Sri Lanka and Uganda; see also case study for Uganda in M. McCallin: *The reintegration of young ex-combatants into civilian life* (ILO, Geneva, 1995)
[211] Case study for Cambodia
[212] Case study for Burma/Myanmar

National Guard, the National Police and the Treasury Police, minors generally did not bear arms but performed other tasks, for instance, serving as "ears", informers, messengers etc.[213]

... children were used by the [opposition] force to run espionage and infiltration missions. Children of tender age were used to carry out these missions.[214]

... the smaller male child soldiers are used to carry out household chores, information gathering/intelligence, running of errands and setting up of ambushes. In this case, the child soldiers are posted on highways and strategic places under the pretext that they are desperately in need of help/assistance. When the opposing forces come in contact with them, they in turn signal to their group for attack and in the process they (child soldiers) are sometimes killed.[215]

Not surprisingly, where the recruitment of children is prevalent, including their employment as spies, this tends to have a detrimental effect on the safety of all children in the conflict area:

[Government] soldiers armed with knives, machine guns, rifles and grenades assassinated 69 peasants, among them 21 children under 5 and another ten between five and ten. An interview is quoted with Sub-Lieutenant ... in which he affirms that even the children were "dangerous", because ... "guerillas begin to indoctrinate them at two, three or four years, getting them to carry things and taking them to different places."[216]

The deliberate killing of children in this way, as well as the repeated references to children being killed by government armed forces when trying to evade recruitment, during training and military service and on capture, raises serious questions about government compliance with the require-

---

[213] Case study for El Salvador
[214] Case study for Ethiopia
[215] Case study for Liberia
[216] Amnesty International Report quoted in case study for Peru (translated from the Spanish)

ments of international human rights law. (Leaving aside the questions of international humanitarian law). Even if they are alleged to have committed an offence, the age of criminal responsibility needs to be taken into account, and they are entitled to the rights of due process. In any case those under 18 years of age at the time of the commission of the offence should not be executed.[217]

Size and agility (and greater expendability) also mean that children are given particularly hazardous assignments:

> *"I was in the front lines the whole time I was with the [opposition force]. I used to be assigned to plant mines in areas the enemy passed through. They used us for reconnaissance and other things like that because if you're a child the enemy doesn't notice you much; nor do the villagers."*

> *"... when I was sixteen. My job was to run out into the "Killing Fields" [no man's land] and grab weapons, watches, wallets and any ammunition from the dead soldiers, and bring it back to the bunkers ... This was a difficult job as you could see the enemy and they could easily "pick you off" as you ran out and back again. The major and I had to do this job as we were the smallest."[218]*

> *Children are used to detect explosive weapons (bombs, claymore mines, etc) ...[219]*

> *Young soldiers normally stayed in front of the others and they were normally used to demine, or to observe the situation.[220]*

Suicide missions are considered to be the particular province of adolescents:

> *They are more physically suited for carrying out such operations and are better disposed to it mentally.[221]*

[217] International Covenant on Civil and Political Rights, Article 6(5), a non-derogable right; Convention on the Rights of the Child, Article 37(a)

[218] Both testimonies of opposition child soldiers quoted in the case study for Burma/Myanmar

[219] Case study for Guatemala (translated from the Spanish)

[220] Case study for Cambodia

[221] Case study for Lebanon (translated from the French)

Urban volunteers act as "instigators, agitators, defenders of the revolutionary principles" as well as participating in acts of sabotage.[222]

However, children are used in the front lines either routinely or if required for a large-scale battle.[223] Inexperience and lack of training results in a high number of casualties: "these children are massacred like flies".[224]

One commander observed:

"When there is shelling, the younger ones forget to take cover. They get too excited. They have to be ordered to get down inside the bunkers."[225]

A journalist, having witnessed the death of an 11-year-old recruit who was "virtually cut in half by a 120 mm mortar explosion" noted:

"He did not duck. He didn't seem to understand the ferocity of these weapons."[226]

There are reports of large numbers of child soldiers, often plied with alcohol and drugs, being used in human wave attacks:

... there were a lot of boys rushing in the field, screaming like banshees when they rushed the barbed wire (barricade in the field). It seemed at first like they were immortal or impervious or something, because we shot at them but they just kept coming.[227]

... in a major battle ... children were used for a massed frontal attack.[228]

A number of child soldiers are forced to commit atrocities against their own family or community in order to cut off the possibility of return,[229] or

---

[222] Case study for Guatemala

[223] Case studies for Afghanistan, Burma/Myanmar, Burundi, Cambodia, Chechnya, Colombia, El Salvador, Ethiopia, Guatemala, Liberia, Mozambique, Turkey and Uganda

[224] Case study for Burundi (translated from the French)

[225] Case study for Burma/Myanmar

[226] Case study for Burma/Myanmar

[227] Case study for Burma/Myanmar

[228] Case study for Sri Lanka

[229] Case studies for Afghanistan, Bhutan, Burma/Myanmar, Colombia, Guatemala, Honduras, Mozambique and Nicaragua

to participate in operations against the peasant community, where many were born and raised:

> There were some who took part in clean-up operations carried out by [the government forces] in the poor districts of the country's cities and in the rural zones. These consisted in searching houses suspected of sheltering [opposition] guerillas, arrests, beatings, imprisonments, tortures and, in not a few cases, the disappearance of young people suspected of carrying out subversive activities.[230]

One report cites the use of under-age soldiers to undertake labour, under the same disciplinary military regime and without payment, in the homes and businesses of military officers.[231]

In urban situations, children are involved in street resistance, riots, throwing petrol bombs or stones at law enforcement personnel, and so on.[232]

A particular task assigned to children may be the recruitment of other children, whether by persuasion, threats or kidnapping by force.[233]

PUNISHMENT

It is not always possible to distinguish routine treatment from specific punishment. There are frequent reports of beatings by superiors, simply as random occurrences but sometimes as punishment:

> Sometimes when I fell asleep when I was on sentry duty, I was beaten by my corporal. He beat me like a dog, like I was an animal, not a human being. There were two or three suicides during that time; of boys who had been hospitalised and finally shot themselves.[234]

Some discipline the children by means of verbal reprimands or the assignment of practical tasks, such as digging holes for rubbish, cutting wood and

[230] Case study for Nicaragua (translated from the Spanish)
[231] Case study for Paraguay
[232] Case studies for the Intifada, Northern Ireland and South Africa
[233] Case study for Peru
[234] Case study for Burma/Myanmar

giving courses in politics to new recruits. Others dole out extra physical training with the intensity varying in accordance with the gravity of the offence, for example, push-ups for something less serious, two hours or many days of forced exercise for a more serious offence:[235]

> *... one 15-year-old soldier ... was playful and often irresponsible. Due to his negligence during duty hours he was regularly punished (front/back-roll on the ground or carrying sandbags for several hours).[236]*

Many child soldiers receive severe physical abuse for failing at military training, falling asleep on sentry duty or disobedience.[237] In one situation child soldiers were reported to receive 150 lashes on the back, neck, head and legs for "lesser crimes".[238]

> *As soldiers in the [opposition] bases, the children were severely punished if they disobeyed the orders of their superiors. The punishments varied from simple corporal punishment, deprivation of food, amputation of fingers, nose, ears, and even execution carried out by one of the more "mature" children.[239]*

> *... the "Tripod" which consists of supporting the body on the tips of the toes, with the hands behind and the head resting on some uneven surface (gravel, bottle tops, a paving stone, etc) for as long as the superiors order.[240]*

In government armed forces, military law applies to all soldiers. The implications of this for child soldiers deserve further consideration.

For attempted escape and disobedience children are beaten to death or shot. Many case studies report children being shot for trying to escape from recruitment or afterwards:[241]

---

[235] Case studies for Colombia, El Salvador, Paraguay and the Philippines
[236] Case study for Bhutan
[237] Case studies for Burma/Myanmar, Honduras and Paraguay
[238] Case study for Uganda
[239] Case study for Mozambique (translated from the Portuguese)
[240] Case study for Honduras (translated from the Spanish)
[241] Case studies for Colombia, Ethiopia, Liberia and Uganda

Soldiers were regularly shot, accused of wounding themselves in order to be hospitalised, retreating in the face of the enemy, or simply gossiping.[242]

Generally, punishment is to be sent to the front line or to have salary withheld but discipline in the army is so poor that any attempt to impose sanctions or punishment is met by desertion or by violence. Reports in the [press] during the 1994 offensive ... featured chilling accounts of out-of-control troops killing their own officers.[243]

In at least one opposition group, members are shot for stealing.[244] Even where death is not inflicted directly it figures frequently as the indirect consequence of other punishments. The repeated references to the killing of children at all stages in the process raise the question of how many of those tapped for recruitment ever make it to the front line, let alone survive to be demobilised.

DEATHS, INJURIES AND HEALTH CARE

... young soldiers are more likely to get killed or injured than adults because they are braver, they tend not to know their locations as well as adults, have often been less well trained or simply have less survival experience.[245]

In most cases, children receive the same training and health care as other soldiers. What this means depends on the resources available and the state of the conflict at any given time. New recruits who are injured, fatigued or simply hungry are often shot and left to rot.[246] Sometimes, both government and opposition groups, during extremely heavy fighting, simply leave the wounded on the battlefield.[247] At other times, badly wounded soldiers are

[242] Case study for Ethiopia
[243] Case study for Cambodia
[244] Case study for Colombia
[245] Case study for Cambodia
[246] Case studies for Burundi and Uganda
[247] Case study for Guatemala

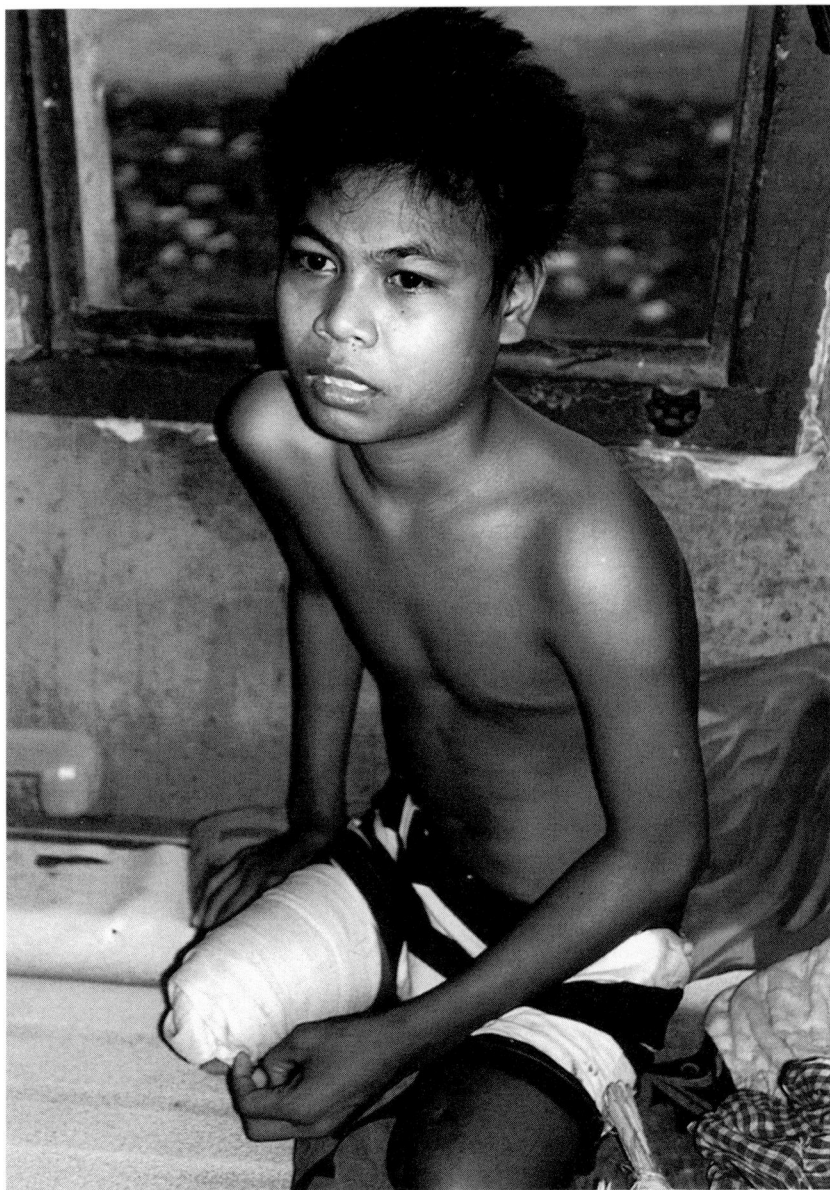

This Cambodian boy joined the army at the age of 15. He fought the Khmer Rouge for four years, until he stepped on a landmine in 1995 and lost both legs.
Photo: Carl von Essen.

"just killed".[248] On the other hand, they may be treated by doctors,[249] at military hospitals, or in barracks by the "medical team".[250] Sometimes, armed opposition groups send injured children home to their families or to foster families to be looked after, or they are taken to the outskirts of settlements so that they can be taken in by the local community.[251] At times, the only medicine available is herbal.[252]

> Through a lack of medical services up to the task of meeting the vast needs, child soldiers who were injured, mainly through mines, and shrapnel from other explosives, ended up losing limbs which were amputated, thus not receiving the necessary medical care.
>
> ... many [demobilised child soldiers] complained of health problems related to the presence of foreign bodies in their organs (bullets, shrapnel ...). This is a problem as many families do not have the resources to pay for operations to remove these objects.[253]

There are also reports of medical supplies being reserved for officers or stolen by superiors, either to be sold and replaced by cheaper supplies or for personal profit.[254] In other cases, financial and other care is provided for some who are seriously injured in battle, as well as for families of war victims.[255]

The most frequent references to specific injuries suffered by child soldiers are to loss of hearing, loss of limbs and blindness:[256]

> The main injuries received by the children are deafness, blindness, burns, damaged limbs leading to amputation, given the frail nature of

[248] Case studies for Burma/Myanmar and Cambodia

[249] Case studies for Colombia and Sri Lanka

[250] Case studies for Cambodia, El Salvador, Ethiopia, Guatemala, the Philippines, Turkey and Uganda. The qualifications of the medics vary significantly.

[251] Case studies for El Salvador, Guatemala, Mozambique and the Philippines

[252] Case studies for Mozambique and Uganda

[253] Both from the case study for Mozambique (translated from the Portuguese)

[254] Case study for Burma/Myanmar

[255] Case studies for Guatemala and Turkey

[256] Case studies for Afghanistan, Burma/Myanmar, Cambodia, El Salvador, Guatemala, Lebanon, Mozambique and Uganda

their bodies, and from the hazards of carrying heavy weights, inhaling toxic substances, land mines, and long marches, to name but a few.[257]

In a direct interview with an ex-soldier, the principal causes of death and injury of minors enrolled in the army were stated to be the explosion of claymore, anti-personnel and anti-tank mines placed by the guerillas. This situation is due, according to the ex-soldier, to the use of minors as advance scouts and as mine detectors. Other causes of death and injury among minors were the explosion of grenades, rockets and bombs. These explosions cause serious injuries and permanent damage, such as the mutilation of limbs, deafness, blindness, total or partial paralysis and on other occasions death.[258]

[A 16 year old] was involved in all the battles and very soon lost his hearing due to his having fired a B7 cannon (105 mm without recoil) non-stop during a battle.[259]

Mine injuries remain among the most common. The other injuries are from the shrapnel of explosions from shells, mortars, B-40s as well as bullet wounds. Losing a foot or a whole leg or both legs from landmine explosions is common. The loss of arms and hands, lacerations to the body, loss of eyesight and hearing are also common injuries among soldiers.[260]

There are conflicting claims about which children are most at risk of death or injury:

Such injuries were most prevalent in the inexperienced child soldiers, and, therefore, the most affected are boys between 11 and 15.

Some suggest that it is the youngest and most inexperienced. Others suggest that these are kept behind the lines and therefore those in the next age

[257] Case study for Mozambique (translated from the Portuguese)
[258] Case study for Guatemala (translated from the Spanish)
[259] Case study for Lebanon (translated from the French)
[260] Case study for Cambodia

group are most vulnerable. In all cases, the children were considered to be at greater risk than adult and experienced soldiers:

> *Carelessness, insufficient training and over-excitation were the principal causes of the increased death toll amongst child soldiers.*[261]

> *There were a lot of injuries due to: (a) fighting or quarrelling between themselves, (b) battlefield injuries and (c) drowning when crossing rivers while on patrol (the young, small boys could not wade in the deep water like the adults while carrying heavy materials like packs or guns). In his battalion, of 30 child soldiers, 15 died or were killed.*[262]

In any event, age is relative and all those falling into the above categories are still children whose bodies have not fully developed. Consideration therefore needs to be given to the additional emotional, psychological, physical, educational, economic and social disadvantages entailed as compared to adults. Loss of sight or hearing may be severe impediments to future educational, vocational or social development. Loss of limbs may sometimes entail repeated amputations for those still growing since the bone of the amputation site grows more than the surrounding tissue, and they will also require new prostheses frequently.[263] In addition to the trauma involved, costs may be too high or the necessary facilities may be unavailable. In societies with high levels of unemployment, the additional disadvantages of such handicaps may be the last straw (see Chapter 5).

TREATMENT ON CAPTURE

> *It so happens that in many instances, under-age suspects never do reach [the] courts and are often nothing more than the usual daily news headlines on [government] television: "Troops have killed so many terrorists today."*[264]

---

[261] Case study for Afghanistan
[262] Case study for Cambodia
[263] S. Roberts and J. Williams: *After the Guns Fall Silent* (Vietnam Veterans of America Foundation, Washington DC, 1995), p 10
[264] Case study for Turkey

Governmental treatment of captured child soldiers varies enormously. Many are treated the same as captured adult soldiers which, in the context of the situations considered for this report, means that they may be treated as criminals or terrorists[265] or they may be held in military prisons.[266] Many captured child soldiers (of both sexes) are subjected to abusive interrogation procedures, torture, taunts, isolation, rape with various implements, being bound in the "banana" position, promises of rehabilitation or threats of death, and being held for prolonged periods of detention or imprisonment. Some die as a result.[267]

Some government armies and armed opposition groups retrain captured child soldiers and incorporate them into their own ranks or give this as an option.[268] This is a distinctive feature of the lives of child soldiers as opposed to adult soldiers and presumably reflects the greater acceptance of participation as normal or inevitable by the children themselves and of the commanders' belief that the children can be retrained to participate on the opposite side.

The fate of those caught spying, however young, is usually death, sometimes accompanied with torture, even if other captured child soldiers are treated as prisoners of war "and render duties such as digging trenches and logistic services".[269]

The inadequacy of detention facilities and lack of adherence to the concept of due process affect children and adults alike but, in particular, there are rarely special facilities and protection for children in such circumstances,[270] although this is required by the Convention on the Rights of the Child, the International Covenant on Civil and Political Rights and other

[265] Case studies for Lebanon, the Philippines, Rwanda and Turkey
[266] Case study for Lebanon
[267] Case studies for El Salvador, Ethiopia, the Intifada, Guatemala and Turkey
[268] Case studies for Afghanistan, Cambodia, Ethiopia and Mozambique
[269] Case study for Afghanistan
[270] Case studies for the Intifada, Lebanon, the Philippines, Rwanda and Turkey; see also Amnesty International, "Sierra Leone: Prisoners of war? Children detained in barracks and prison" (12 August 1993)

instruments.[271] Long delays in the administration of justice impact severely on children because of the deprivation of family or substitute care, education, poor food and health provision, and so on. This is one of the reasons why Article 37 of the Convention on the Rights of the Child provides: "The arrest, detention or imprisonment of a child shall be used only as a measure of last resort and for the shortest appropriate period of time." Social welfare institutions may be reluctant to take them on fearing that they will be a bad influence on other children being accommodated.[272]

Detention and questioning of children also occurs when there is no specific evidence of their participation in the armed conflict:

> ... the army's counter-insurgency operations involve the systematic detention of men, women and children found in areas recaptured from rebels. Young people, children and babies have been caught up in the army's efforts to identify rebels and those who may have helped or collaborated with them.[273]

In addition to the random detention of children as part of the civilian population in this way, where children have been recruited into armed opposition groups, those in the relevant age group may be detained:

> Often, the only grounds for detention appeared to be a general suspicion that men and boys of fighting age may have fought with the rebels.[274]

This illustrates the problem that, as long as children are recruited into armed forces or armed opposition groups, they themselves and other children cannot be insulated from the consequences of, or the suspicion of, involvement.

Age was a consideration in the minds of some captors. Some have a pol-

---

[271] Convention on the Rights of the Child Articles 37 and 40, International Covenant on Civil and Political Rights Articles 9, 10, 14, and 6 (in relation to the death penalty), and the UN Standard Minimum Rules for the Administration of Juvenile Justice (known as "The Beijing Rules")

[272] Case study for the Philippines

[273] Amnesty International: "Sierra Leone: Prisoners of war?" (12 August 1993)

[274] ibid

icy of treating child soldiers as innocent victims: trying to scare off adult soldiers in order to gather up the children without having to fight them, questioning them for intelligence and to ascertain information necessary for family reunification, and then handing them over to others for processing, although the policy may not always be implemented in practice.[275] One interviewee noted that some minors captured by the government forces were not killed and concluded that this may have been because of their age as no adult soldiers were "saved".[276] There is always a danger that governments will use captured child soldiers for propaganda purposes.[277]

It may only be when captured and charged that the real age of the soldier becomes known, either because this is the first time proof of age is sought[278] and/or because there is an incentive for the accused to reveal it when faced with criminal proceedings:

> *Many instances arose where youths (under 17) lied about their age to get into the [armed opposition group] – in lots of cases their true age only became known when they were charged.*[279]

> *... a soldier who accidentally killed his army friend and who, when brought before a military court, did not hesitate to present his real age so that his case would be transferred to the juvenile court where punishments are automatically reduced.*[280]

One of the consequences of child recruitment into armed opposition groups can be that the age of criminal responsibility is reduced and/or that anti-terrorist or emergency legislation is introduced which takes no account of the age of suspects, or reduces the age of applicability below that for "normal" criminal offences:[281]

---

[275] Case studies for Mozambique and Uganda
[276] Case study for El Salvador
[277] Case study for Turkey
[278] Case study for Guatemala
[279] Case study for Northern Ireland
[280] Case study for Lebanon (translated from the French)
[281] Case studies for the Intifada, Nicaragua, Peru and Turkey

Children were regularly interrogated using the same physical and psychological pressures that are applied to adults.[282]

Although this may be seen as a normal reaction for a government faced with this particular problem, it raises questions about compliance with the provisions of the Convention on the Rights of the Child, and of the adequacy of protection and safeguards for child suspects, such as access to a lawyer, suitable rehabilitative provision for those found guilty and the non-application of the death penalty to those under the age of 18 at the time of the commission of the offence. The question of the responsibility of child soldiers for their actions, the age of criminal responsibility, appropriate determination procedures and disposal arrangements need further consideration.

[282] Case study for the Intifada

CHAPTER 5

Demobilisation, Rehabilitation and Social Reintegration

THE PURPOSE OF THE Child Soldiers Research Project was to develop a better understanding of both the causes and consequences of children's participation in armed conflict. It is at the time of demobilisation, and the children's return to their families and communities, that the profound effects of their participation become apparent. The information from the research indicates that the children will confront a situation of multiple risks which can significantly compromise their reintegration into civil society. This situation is a function of the events they experience as soldiers, as described above, but is also due to the fact that the living conditions of their families and communities, which in most cases occasioned their participation in the first place, are even more precarious than when the children were recruited.

In outlining various issues that were raised in the case studies, this chapter discusses the consequences for the children as a continuum that includes their experience before and after recruitment, not only their experiences as soldiers. We believe this is important, firstly to emphasise from the perspective of the children themselves how their experiences can affect their developmental processes, and thereby constitute a gross infringement of their rights within the terms of the UN Convention on the Rights of the Child (CRC).

In addition, it is only through an appreciation of the conditions of life which define the children's social and cultural environment upon demobilisation that effective strategies can be implemented to address the consequences of their participation, and effect their social reintegration. Such a framework incorporates attention to the reasons for the recruitment of children, as much as the consequences, through action which enables families and communities to protect their children. This is a theme that was emphasised in the case studies for the research, where repeated reference was made to the linkage between education, employment opportunities, and the economic security of the children's families as the factors that would not only determine successful social reintegration, but also contribute to efforts to prevent further recruitment.

DEMOBILISATION

The first stage in this process is obviously to secure the demobilisation of the children, and to ensure that this occurs in such a manner that programmes can be implemented to address their special needs. The information on the demobilisation of children provided for the research project in fact describes only limited instances where child soldiers have been demobilised, and where some effort was made to assist in their rehabilitation and reintegration.

The only example of an official demobilisation of children was described in the case study for Liberia. A small number of children (184, or 6% of the total number of combatants disarmed at that time) were included in a process of planned disarmament and demobilisation of combatants in March and April of 1994. The climate for demobilisation resulted from the signing of the Cotonou Agreement in July 1993, though this did not specifically include recognition of the role played by children in the conflict, nor specify strategies to assist in their rehabilitation and social reintegration. The emergence of factions who had not been signatories to this agreement, and the subsequent resumption of hostilities "complicated the already fragile demobilisation... and brought the process to a complete halt."[283]

Starting in March 1994 UNOMIL,[284] through its spontaneous demobil-

[283] Case study for Liberia
[284] United Nations Mission in Liberia

112

isation programme, demobilised a further 434 child soldiers aged 17 years and under. It would appear that almost all of these children were demobilised with an adult relative who had been serving with the same forces. This characteristic of service in Liberia has resulted in children returning directly to their communities of origin, with only a small number expressing a wish to take advantage of the programmes of rehabilitation that are available.[285] Subsequent demobilisation of children in any planned manner was considered to be dependent on the implementation of procedures for general disarmament and demobilisation negotiated through "the interim government ... local and international relief organisations, civil/interest groups, and church-related organisations."[286] [287]

Although not included as a case study for the research project, an official demobilisation of children during an on-going conflict occurred in Sierra Leone, which was followed by a programme to address their rehabilitation and social reintegration. In June 1993, in recognition of their obligations following ratification of the CRC, the government army demobilised 370 children. It should be noted that this also resulted from advocacy on the part of the local UNICEF office to release the children.

The case study for the Philippines describes demobilisation of children during conflict when they become "bored or demoralized" or when there

[285] There are also reports that in Liberia whole families move with certain factions. A similar involvement of families is described with the Mujahideen in Afghanistan. Whether such involvement reduces the effects of participation on children, and/or contributes to effective rehabilitation and social reintegration, is a matter for further enquiry. Some children with the PKK may also be with their families, as "most had been flushed out of their homes along with their families and had accompanied them to the camps where they joined the *freedom movement*." (Case study for Turkey)

[286] Case study for Liberia

[287] In April 1996 fighting again broke out in Monrovia. At the time of finalising this chapter (June 1996) there was no peace initiative and, rather than progress in the demobilisation of children, it was considered more likely that there would be further recruitment of under-age combatants, including re-recruitment of children who had already been demobilised.

has been a "violation of disciplinary regulations."[288] This may be due to the "special consideration" given to these children who were serving with the NPA, where there is monitoring of their well-being. A contrast to this example is given in the case study for Sri Lanka, where demobilisation of children with the LTTE is not only impossible, but desertion also is prevented and punished.

> *Once recruited, a child cannot leave the LTTE. Their hair is cropped as soon as they are enlisted. Some deserters hide till their hair has grown so they cannot be discovered. Others who wanted to go back to their parents are given a sound thrashing before their parents and sometimes by their peers.*[289]

In Afghanistan "there is no planned demobilisation scheme to assist and facilitate the return of child soldiers to civilian life." In this instance, demobilisation of children was and is a function of the particular group with whom they were serving. Children who served in the forces of the pro-communist regime were required to serve for four years and were then demobilised, although no mention was made of any financial or other material benefits awarded at this time. They could also be released from service as a result of a disability – presumably as the result of injury during conflict. Desertion of children was a common phenomenon, and was prevented by sending them "to unfamiliar, remote areas far away from their homes." Children who fought with the Mujahideen between 1978 and 1992 could, in principle, leave whenever they wished. There was, however, encouragement for them to remain both from religious persons and their families. As in this instance the children's families were present with them, it would have been difficult for them to resist the social pressure that would seem to have been present. There is no specified period of service for children presently fighting with different warring factions (1992 to the present). The

---

[288] Also in the Philippines' case study there is reference to a rebel returnee programme for those who apply for the governmental amnesty, but there is no record in the case study of anyone receiving assistance under the terms of this programme.

[289] Case study for Sri Lanka

case writers in this instance consider that these children "have no option left except to stay in the army" as there is a lack of facilities, the children are poorly educated, and the army "at least provides them with earnings."

A disturbing form of demobilisation, and certainly not one to be advocated, is described in the case study for Colombia where, if a young recruit "manages to kill a 'subversive', he is demobilised and returned to civilian life, 'as a form of payment'."

In Burma/Myanmar, the case writers had "never encountered evidence of an official demobilisation procedure for child soldiers working with the Tatmadaw" and, despite "signings of a number of cease-fires, there has been no significant demobilisation of either government or opposition troops." As "troop numbers are made up mainly of very young people", the process of demobilisation here would seem to require not only recognition of the participation of children, but special considerations for rehabilitation and reintegration. It is not considered that this will happen as there is a denial that children are involved in the military.

In on-going conflicts there is little hope of demobilisation, even when this may have been specified in peace agreements, such as in the case of Guatemala. Here army conscripts are demobilised at the end of their term of service, if they have infringed military law,[290] or if they are disabled. The case study for Guatemala did also mention that under-age conscripts can be demobilised if they have proof of their age, but the numbers were not specified. The problem here may be that many children from impoverished or marginalised communities will not be in possession of the appropriate documents. No form of demobilisation is available for those serving with the guerilla movement. Their service ends with capture by the army, or

---

[290] Such an infringement may lead to prosecution and imprisonment, and thus also has implications for the legal process to be followed where the offender is under age. A related concern is noted in the case study for Peru where "adolescents supposedly responsible for the crime of terrorism could be condemned to various sentences, including life imprisonment." A fuller investigation of how child combatants are treated within the legal system is required, particularly the extent to which the system may be used to exact 'retribution' on the part of one faction.

through desertion or disability. A similar situation pertains in Turkey for children serving with the PKK. Their term of service is not specified and they remain until "they are killed or assigned at a later stage to peacetime activities (i.e. activities in Europe, publishing, propaganda etc.)."[291] The latter could be viewed as a form of demobilisation, but is obviously based on some form of selective process, and raises the question of *who* is considered suitable for these activities.

Certain of the case studies described forced conscription of under-age recruits to government forces where there is a set term of military service. Whether such under-age conscripts are in fact released at the end of their term of service is a question that needs further exploration. One case study stated that they do in fact "remain in military service longer than the time prescribed by the law (2 years), and have to claim or demand their release, otherwise no one remembers that the period has been served."[292] In El Salvador and Afghanistan, although the period of military service may have been extended during the conflict – in the latter case from two to four years under the pro-communist regime – release did take place upon completion of service.

The demobilisation of children during conflict is influenced by the reasons for their recruitment, whatever the manner in which this is conducted. Advocating for their demobilisation will in some cases confront complex social, political and cultural issues that are directly associated with the causes of the conflict, and/or the manner in which it is prosecuted. Where people "cannot escape" from a conflict that "is taking place in their communities and in their own homes",[293] they are unable to protect their children from recruitment and even less likely to be able to effect their demobilisation. This situation may be further compounded by cultural conceptions of appropriate roles and duties of children, particularly when specific groups or communities are targeted, and even how "childhood" is itself determined. The case study for Chechnya describes such an example, where

[291] Case study for Turkey
[292] Case study for Honduras
[293] Case study for the Philippines

youths between the ages of 16 and 18 were not considered children" and thus their participation is considered "normal" and "natural". Participation of some children in the Lebanon is sanctioned by religious precepts "in accordance with which Jihad becomes an obligation for the religious faithful including children if they have reached the age of majority according to the chari'a, i.e. if they have reached puberty."[294]

The information from the case study for Afghanistan provides very clear examples of how demobilisation is a function of the form of recruitment and of which faction is responsible. In some cases the supportive attitude of their families and communities towards their participation[295] may prolong their involvement. Ideological commitment can be a significant factor in convincing people of the legitimacy of involving children, and even encouraging their participation. This is a particularly difficult context within which to introduce concepts of child rights and welfare as the basis for advocating for prevention of recruitment and demobilisation. It was an issue noted in certain case studies, but not discussed in any detail, and is certainly one that requires further investigation.

> ... a spirit of solidarity developed amongst the people to voluntarily join the Mujahideen (resistance) forces and offer the services of their young children. This was done mainly based on their religious beliefs.[296]

> For the MNLF, child soldiers are a natural outcome of their holy war ... Muslim male children are reared side by side with a gun. At nine they can dismantle and put back an armalite. (A TV) news feature on the MNLF shows families socializing children into the holy war from the time they are born. Especially if fathers or family members have been killed in the conflict. For them, the holy war is a noble commitment and a natural consequence of their religious beliefs and cultures.[297]

Tacit approval, if not direct encouragement, for the involvement of children in conflict may also characterise the child's social environment. This will be

---

[294] Case study for the Lebanon
[295] Palestine; Philippines; Chechnya; El Salvador; Colombia
[296] Case study for Afghanistan
[297] Case study for the Philippines

the case particularly where there is glorification of the military as, for example, in Burma/Myanmar.

> *Children, especially young boys, are raised to revere military leaders of the past, and to look on military induction as a sign of manhood. In much of the popular media, the soldier is held up as the perfect role model ... To be a soldier is to occupy a position of great honour and self-sacrifice. The emotional pull of such prestige should not be underestimated.*[298]

It will be apparent that in such situations advocating for the demobilisation of children may be seen as a direct challenge to cultural beliefs.

People's living conditions may also be a significant factor that not only influences recruitment, but prevents the demobilisation of the children. From the perspective of those involved, such situations may result from haphazard conditions, as the conflict moves from one location to another, or from the ways in which the conflict affects the capacity of different groups to protect their children.

> *The means of protection that families have developed to keep their children out of the war ... depend on the different phases of the war and on the situation that a family found itself in: what was applicable to Lebanese was not applicable to families from ethnic or religious minorities or to refugees due to the fact that they do not have roots in the Lebanon and therefore no extended family or home village. (As the war progressed) impoverished families resigned to their fate saw the engagement of their sons in the militia as a means of protection, a last resort in the absence of any alternatives."*[299]

> *The need for self-defence as the reason for involvement in armed forces characterises those cases of child soldiers who lived near the front line and whose lives and the lives of their families were in danger because of the conflict. Sometimes these boys were the only male members of the*

---

[298] Case study for Burma/Myanmar
[299] Case study for Lebanon

family as the adult men were serving as soldiers in some other armed
groups. This was mostly the situation in rural settings.[300]

Families may even see material advantage, regardless of any ideological commitment, in their children's involvement, and be reluctant to forego the benefits that the child combatants obtain for their families.

For many families allowing their children or volunteering to send their
children to fight for the country was a good thing. According to CAW
field experience many mothers have remarked on the joy of seeing their
ten-year-old dressed in a brand new military attire carrying an AK-47.
For some families the looted property that child soldiers brought home
further convinced them of the need to send more children to the war
front to augment scarce income.[301]

Even in situations where communities *in principle* may not approve of children fighting, in some way these attitudes and beliefs are set aside in time of war, when they do "not have time to give much thought to these things".[302] Additionally, some children may join armed groups for protection because they have "lost both parents in the war ... and had no-one else to take care of them".[303] In such situations, demobilisation is virtually impossible without substitute care being provided – an unlikely event in the midst of conflict – and, by association and identification with the adult soldiers, children may be drawn progressively into assuming an active role in the conflict. When the children's primary source of protection is thus denied them, their situation poses a high degree of risk. It will be apparent, therefore, that in cases where factions forcibly recruit children, and in addition deny their participation, there is little likelihood that they will be

---

[300] Case study for the the former Yugoslavia

[301] From the case study for Sierra Leone prepared by the project Children Associated with the War (CAW) in: M. McCallin *The reintegration of young ex-combatants into civilian life.* ILO, Geneva (1995)

[302] Case study for El Salvador

[303] Case study for El Salvador. Note also the situation in Rwanda where some 5,000 children are being cared for by the RPA following the genocide of 1994. Here the military emphasised its protective role as the child's substitute family.

prepared to demobilize such a readily available source of manpower for any reasons associated with their welfare.

For those case studies from countries where conflict has ended, the situation is little better for the children. In no situation has there been formal recognition in peace agreements of the participation of children, even when their involvement has been extensively documented. In Mozambique, for example, where Renamo's recruitment of children was well known, "public recognition that children had been actively involved in the conflict was considered a political liability, and their demobilisation was inconvenient".[304] Although there has been no formal denial of the participation of child soldiers in Ethiopia, there have been no special efforts to address their situation in the demobilisation process. The large numbers of children and young people who were involved may have defied any attempt to target them as a matter of priority, although the education of some 8,500 is reportedly the responsibility of the Ministry of Defence.

In Cambodia, where again there has been no denial of the participation of child soldiers in the conflict, "there was no specific mention of child soldiers in the Paris Peace Accords of 1991. The demobilisation survey and proposal in 1992 does not deal with the problem or the specific issues surrounding possible demobilisation of child soldiers. It is implicit that they are seen in the same light as adults."[305] The proposal for the demobilisation of some 43,000 soldiers between 1996 and 1998, sponsored by the World Bank, includes no provision for child soldiers.

In South Africa, although there has been no formal demobilisation of young people involved in the self-defence/protection units, the normalisation of their daily lives is being addressed through a programme of re-education, retraining and vocational programmes, which is the responsibility of the Ministry of Welfare and Population Development. This process should be assisted by the process of disarmament which the government has implemented "by declaring an amnesty wherein all members of former units protecting communities could hand in arms in their possession".[306]

[304] M. McCallin (1995). *The reintegration of young ex-combatants into civilian life.* Report prepared for the International Labour Office, Geneva.
[305] Case study for Cambodia
[306] Case study for South Africa

In some instances[307] it was noted that material benefits in the form of cash, rations or clothing were awarded to former child combatants. In others, limited initiatives on the part of the government[308] and NGOs and other local organisations[309] are being implemented which seek to address the children's needs, notably education, vocational training and psychological counselling. Attention to former child combatants may also be integrated into programmes for children affected by war, such as the programme of family tracing implemented by UNICEF, Rädda Barnen and the ICRC for Sudanese unaccompanied minors, amongst whom were a significant number who had been directly involved in the fighting.

Whilst every effort must be made to secure their demobilisation, it should not be regarded as a discrete event, an end in itself, but more the initial step in rehabilitating and reintegrating the children. The fact that children's rights and welfare are either marginalised or completely ignored, even in official procedures for demobilisation, highlights the extreme vulnerability of children who are not only easily recruited, but can be summarily demobilised without due regard to their special needs. Thus the continued inadequacies of the international system in protecting their rights, and securing their well-being following demobilisation, must be emphasized.

These inadequacies are disturbing, given the emphasis many case-writers placed on the manner in which children have been affected. The experiences and events that characterize their involvement mean that, in a context of violent separation from their families and communities, they have lost their childhood and chance for education, whilst being "indoctrinated to obey orders without question" and "programmed to solve problems using a gun".[310] Recognition of their participation and its consequences for

[307] Burma/Myanmar; Ethiopia; Liberia
[308] a) Mozambique: Programme of Support for Children with War Experience implemented in 1995 by the Ministry for the Co-ordination of Social Action.
b) South Africa: literacy and counselling programmes, and assistance with finding employment for 13–18 year olds involved in the SDU
[309] a) Liberia: programmes of rehabilitation and vocational training
b) Nicaragua: therapeutic centres
[310] Case study for Burma/Myanmar

their development must be premised on a readiness to respond to their needs for rehabilitation and reintegration, and incorporated in peace agreements and related documents. Without this requirement, there is no framework within which planning for the demobilisation of child combatants can be undertaken, or programmes implemented which address their particular concerns. The limited initiatives described above underline the difficulties that can be encountered when such processes are undertaken on an *ad hoc* basis, without the support of policy and planning procedures.

THE CONTEXT FOR REHABILITATION AND REINTEGRATION

The significant rhetoric which has surrounded the phenomenon of the recruitment and participation of children in armed conflict, in particular the sensationalistic focus on "young killers", has hidden the necessity of recognising and addressing the essential child-welfare concerns which characterise every instance in which children are involved in armed conflict. There is almost a sense in some instances in which they are portrayed as being irretrievable; as though the acts they have committed put them outside the boundaries of civil society.

> *Even if they survive the rigors of combat, it's often too late to salvage their lives. Unrelenting warfare transforms them into preadolescent sociopaths, fluent in the language of violence but ignorant of the rudiments of living in a civil society.*[311]

Advocating for the demobilisation of child soldiers, and determining appropriate strategies for their rehabilitation and reintegration requires an understanding of the consequences they suffer *as children*, and a framework within which planning for children can be conceptualised and implemented. Planning for children in such extreme circumstances must incorporate an assessment of the risks to their development that have resulted from participation. It must be complemented by a consideration of the social, political and cultural reality which defines the context for the children's rehabilitation and social reintegration. It must therefore take

[311] Boy Soldiers. Newsweek Special Report, August 7th 1995

account of gender issues, the age of the children involved, and their length of service – a point of particular importance for those recruited at a young age who will have been separated from their families at a time when they "have not yet formed a stable identity which would allow them to psychologically 'survive' under such circumstances."[312]

The implicit needs/rights perspective of the Convention on the Rights of the Child provides a framework for guiding such an assessment through its consideration of the "evolving capacities" of the child, and the provisions of the various articles which ensure the child's right to grow and develop in an environment where her/his needs are met. The applicability of the CRC in the many and varied circumstances described in the case studies is a function of the extent to which it embodies a set of norms on how children should be treated, and represents a consensus on the minimum rights which must be guaranteed to children.

> *The text of the Convention on the Rights of the Child provides an excellent framework for advocacy on behalf of children, regardless of the discipline from which the rights of the child are being approached. It is equally useful to psychologists, educators, religious leaders, lawyers and parents ... Each time the standards of the Convention are invoked in practice, it strengthens its importance in international law.*[313]

Children are involved in an active process of development; they are not static observers of events. Thus whatever their situation, be it as the member of a functioning family, or a member of an armed faction, they continue to develop physically, emotionally and socially. Developmental outcome, or more simply their capacity to function effectively within their societies, is determined by their experiences in this process, be they positive or negative. The cheerful twelve year old who was abducted by an armed group

---

[312] E. Jareg. (1993) *Rehabilitation of Child Soldiers in Mozambique.* Note prepared for the Secretary General of Redd Barna

[313] C. Price-Cohen. *Considerations Affecting the Implementation of the United Nations Convention on the Rights of the Child in Situations of Forced Migration.* In: M. McCallin (Ed.) The Psychological Well-Being of Refugee Children. Research, Practice and Policy Issues. Second Edition (1996). International Catholic Child Bureau, Geneva.

may come home as an aggressive sixteen year old, carrying her own child in her arms, brutalised by abuse and with a confused sense of loyalties and identity.

The CRC, through its emphasis on the child's "best interests", addresses the heart of the issues that must be considered in implementing procedures to secure the well-being of child combatants, and requires us to assess whether and how our interventions meet this requirement.

Planning for the demobilisation of child soldiers must incorporate their special requirements for rehabilitation and social reintegration. It will, therefore, need to take account of the social, political and cultural context that resulted in their recruitment, the reality of their experiences in conflict, and their situation upon return. The experiences and conditions of life of their families and communities are integral components of each child's reality and, thus, efforts will need to be made to raise awareness within communities of the impact of the conflict on their children, and to incorporate their participation into programmes of rehabilitation and reintegration. In itself, this process can make a significant contribution to post-conflict resolution and the prevention of further recruitment. Greater understanding and awareness, and its corollary of greater empowerment, can prove crucial to preventing recruitment, as is evidenced by the disparity between the numbers of children from "educated" families who are recruited, as opposed to those from poor families.

Obviously, the above statements are easily if necessarily stated, whilst the reality is and will be harder to address given the circumstances of recruitment and participation. Several factors will influence the manner in which children's issues are incorporated into demobilisation procedures: the number of children involved, and in which faction; denial of participation/political sensitivity; demands on available resources, particularly in the period of post-conflict reconstruction; needs of other war-affected children. Without efforts to incorporate children's needs *generally* as a priority issue in the process of reconstruction, specific concerns for the situation of child soldiers may be marginalised.

... the State policy regarding the status of children in general remains that they are not "top priority". The government maintains that in the

acute crisis situation which is alive in post-war Lebanon, "children's
needs cannot be a priority".[314]

As with recruitment, attitudes towards the demobilisation, rehabilitation
and reintegration of children may well reflect the low priority accorded to
their well-being. As the case study for El Salvador notes, the hardships suffered during conflict are not necessarily any greater than those imposed by
the poverty which characterises normal life, as evidenced by the ease with
which children from marginalised and impoverished communities are
recruited. In this context, the exploitation and marginalisation of their parents needs to be redressed, a decent standard of living and full protection
of their social, political and human rights afforded them. The effect of such
changes would quickly be reflected in the lives of their children.

In this regard it is worth noting the reservation outlined in the case
study for the Philippines with respect to the term "reintegration".

> *Reintegration to what – a life which is dysfunctional, mainstream soci-*
> *ety that is saddled with social problems? What is indeed normal? The*
> *child soldiers learned poverty and injustices as normal realities of life –*
> *what they discovered though is an ideal when they joined the NPA.*
> *Therefore the use of recovery and not reintegration. For the child*
> *soldiers and rebel returnees, reintegration smacks of surrendering their*
> *principles and ideals to a society which in many ways is plagued with*
> *more problems and may be the very root of these problems.[315]*

This comment implicitly questions the extent to which initiatives directed
at conflict resolution are also concerned to address the root causes of conflict, characterised to a large degree in the case studies by efforts to
address social injustice and the social and economic marginalisation of particular ethnic or minority groups.

The case study for the Philippines also very importantly states that "the
most appropriate policy and intervention can best be identified if the phenomenon of child soldiers is viewed from their eyes, their life situation, their

---

[314] Case study for the Lebanon
[315] Case study for the Philippines

context". With due regard to this comment, the following issues are outlined as of particular concern in planning programmes for the demobilisation, rehabilitation and social reintegration of former child soldiers.

THE FAMILY AND COMMUNITY

> ... the family, as the fundamental group of society and the natural environment for the growth and well-being of all its members and particularly children, should be afforded the necessary protection and assistance so that it can fully assume its responsibilities within the community, (and) that the child, for the full and harmonious development of his or her personality, should grow up in a family environment, in an atmosphere of happiness, love and understanding.[316]

> Children's well-being and development depend very much on the security of family relationships and a predictable environment. War, especially civil war, destroys homes, splinters communities and breaks down trust among people – undermining the very foundation of children's lives.[317]

These two statements describe on the one hand the "ideal" to which all societies should aspire in protecting and caring for their children and, on the other, the "reality" that confronts the millions of children caught up in armed conflict. It is the gap between the two conditions that we seek to close when we consider interventions to address the well-being of children in conflict situations. The child soldiers are certainly amongst the most vulnerable of these children, not least because they have undergone their experiences whilst separated from their families and traditional support structures. Thus the principles of working with unaccompanied/separated children would seem to be those most applicable to their situation.

---

[316] Preamble to the UN Convention on the Rights of the Child.
[317] Promoting Psychosocial Well-Being among Children Affected by Armed Conflict and Displacement: *Principles and Approaches*. Save the Children Alliance (1996). Paper produced as a contribution to the UN Study on the Impact of Armed Conflict on Children.

126

It is not by chance, therefore, that the issues of family and community are given priority in this section. Perhaps the most significant factors influencing the child soldiers' return to civil society are family reunification and reintegration. It is here that the two contrasting aspects of their experience confront one another. The child and the soldier. The child has need of the family, and child soldiers, except in rare situations, have been separated from their families for long periods. Yet the soldier will need assistance to adjust and reintegrate, due largely to the process of "asocialisation" and "total destruction of trust in others" which will have been the experience of the vast majority of these young people.[318] All too often, it is the soldier we consider and we lose sight of the child. In recognition of the crucial role of the family in rehabilitation and social reintegration, planning must ensure that family tracing programmes are implemented as quickly as possible,[319] and children reunited with their families.

This will be an important process for the children, who may be confused by the situation, or indeed reluctant to relinquish their identity as a soldier, particularly when they perceive themselves as having played a role in addressing the ills that their society had experienced. This can result in a sense of abandonment or rejection following demobilisation, or resolution of the conflict. For many of them, the army or armed group will have become "their protector and provider, and they in turn (will have) identified with it. This loss, as much as the direct influence of militarisation, should be recognised as influencing their behaviour"[320] so that social reintegration is based upon a process of re-attachment to their families and communities.

Many children have been deliberately moved far from their families following recruitment.

... many were sent away to remote areas where contact with their home

[318] E. Jareg (1993) op. cit.

[319] For a full discussion of family tracing procedures see: L. Bonnerjea (1994). *Family Tracing. A Good Practice Guide.* Save the Children Development Manual No. 3. London, Save the Children.

[320] McCallin (1995) op. cit.

was difficult, impossible or forbidden. Some observers have speculated that distant field placements are chosen in order to make it difficult for soldiers to desert or defect ...[321]

Due to the process of the conflict families may be displaced, in some cases seeking refuge across a border, or in different areas within their own country. Displacement in such situations may occur several times as the fighting moves from one area to another. In the ensuing chaos, children may be separated from their families and recruited as unaccompanied children, or may join up because of the manner in which the conflict is prosecuted against their families and communities. In such circumstances, family tracing and reunification become very difficult.

> *The conflict in turn has also led to more children being displaced, orphaned or living in families where one or both parent/s are killed or the father is away fighting. Disruption of family units could also motivate a child to join the movement.[322]*

Simply finding and placing the children with their families may not be sufficient, as the impact of the conflict on the family may mean that, on return, the children are confronted by a family that may have radically changed in structure and the manner in which it functions. This may significantly compromise the efforts the child soldiers and their families must make in order for them to return to life in civil society.

> *The simplest form of reunification is the direct return of a child to the members of the family she was living with before the separation. In the simplest reunification she is also returned to the same place they were living in before – displacement of the family was temporary and they are all back together ... the people are the same and the place is the same. But nevertheless the first impression of 'sameness' is illusory: the child has had some very frightening experiences. She has been without protection for a period of time. She may have been beaten, abused, raped. She is in many ways not the same child she was before. And the*

[321] Case study for Burma/Myanmar
[322] Case study for Sri Lanka

family has changed too. Each member has had to develop survival strategies for themselves and for their relatives, sometimes with uncomfortable consequences. They may have had to choose which of their children should live and which should die. The mother may have had to turn to prostitution to survive. The father may have had to betray his community, his tribe, his religion, his caste in order to stay alive ... Reunification is not, therefore, a return to the situation that was before, but rebuilding, recreating a family, with new experiences which the family needs to come to terms with.[323]

In recognition of this, every effort must be made to assist families in this process of rebuilding family life. In Sierra Leone[324] this was accomplished through a programme of Family/Community Sensitisation, Mobilisation and Advocacy. This focused on:

- *promoting reconciliation and reunion among husbands and wives and settling of other family or inter-family or communal disputes; correcting family/community prejudices and misconceptions about child soldiers;*

- *animating and mobilising families and communities into actions on behalf of their children;*

- *creating readiness in families and communities for the return of their children back to the fold, even though times are hard and facilities for the care of children are inadequate;*

- *equipping families and communities with various skills in the handling of ex-combatant children.*

Such approaches have a greater chance of bringing about "a more positive social reality" through building "on a community's own social networks."[325] They are more cost-effective, culturally appropriate and sustainable, and reduce any likelihood that communities will become dependent on external resources to maintain a programme.

[323] L. Bonnerjea. op. cit.

[324] See case study for Sierra Leone in: M. McCallin (1995) op. cit.

[325] *Children in War. Community Strategies for Healing.* Save the Children USA, University of Zimbabwe & Duke University (1995)

There is an awareness that the Government itself will not be able to meet the many needs of the population. Thus there is a need to encourage community initiatives which are usually low cost and of greater impact, as well as to build up local capacity to resolve their own problems through programmes based on the communities themselves, managed by them and that are self-sufficient and sustainable.[326]

This comment also links with the observation made in the case study for the Lebanon that children's needs are not given priority during the phase of post-conflict reconstruction, when human and material resources are limited. Directing these scarce resources to the needs of one group of children – the child soldiers – risks marginalising or stigmatising them, thereby affecting their chances of reintegration. This will be the case particularly where a community blames the children for the events they experienced. Community involvement in their social reintegration would ensure not only the applicability of interventions to the circumstances of the children, but would also be more likely to result in an approach where the needs of the child soldiers were incorporated into overall programming to address children's issues in the period of post-war reconstruction.

Despite programmes of family tracing and reunification, however, there may be some children who will not return to their families. This may be because their families have perished in the conflict, or because they are fearful of returning home as a consequence of the events in which they participated, and may thus remain isolated from any system of care, protection and support. In some instances, even when the conflict has ended, the circumstances of the family may mean that it is "unable to satisfy the many needs of the child; it ends up 'facilitating' the move of the child onto the streets, either through the child's own initiative, or by encouragement or forced by the family to provide for him/herself and even for the family."[327]

For children in such situations there will be a need to consider alternative care-giving arrangements that meet the children's physical, social and emotional needs. This will be of particular importance for the older child

---

[326] Case study for Mozambique
[327] Case study for Mozambique

130

combatants who may have spent considerable time with the military and, as a result, find it very difficult to reintegrate into their families. As certain of the case studies pointed out, this group is in danger of resorting to criminal activities, made easier by the availability of weapons.

The creation of special institutions is not considered a satisfactory solution. They may serve only to marginalise the children even further, and also impede efforts at family tracing and reunification. They are also a costly enterprise which, though they may initially attract funding, are inappropriate in many cultures, and not a sustainable option for countries emerging from conflict. Nor are they the best setting in which to meet the special needs of children who have experienced horrifying and life-threatening events.

> *Children in institutions may get regular meals and exhibit playful behaviour but miss the individual love, nurture, guidance, role-modelling and personal attention that they depend upon for normal growth and development ... Institutionalisation leaves children disadvantaged, like plants that have not received the optimal nutrients for growth. Institution residents are often described as problem children, a judgement, if true, that reflects less about the children than about the failure of the society to meet their needs.*[328]

There is some concern that former child soldiers may need a period of time between demobilisation and return to their families to allow for rehabilitation and assessment of their situation. The acts they committed may have had a serious impact upon them, they may be in poor physical health or have become dependent upon the drugs that were administered to them. What is important to recognise here is that any such programme of rehabilitation does not become an end in itself. It must be seen as a temporary, interim measure. The message must be given to the children and their families that this is not the answer to their situation, but a process which is designed to facilitate their ultimate return to their families and commu-

---

[328] E. Ressler, J.M. Tortorici & A. Marcelino (1993) *Children in War. A Guide to the Provision of Services.* UNICEF Publications.

nities. Such responses will also be a function of the number of child soldiers involved in the conflict, and the availability of resources. The programme in Sierra Leone which implemented such a response was working initially with a small number of 370 children. In countries such as Ethiopia, where the numbers of child soldiers ran into thousands, it would have been an impossible task to implement a programme of this nature. Thus, some thought must be given to this issue when planning the demobilisation of children, in particular preparing communities for the direct return of their children so that rehabilitation can be integrated into community activities.

A more appropriate option may be placement with a substitute family. This "may be a welcome relief to some children, who had been previously fending for themselves" but for others "substitute families may be met by fear, anger and resentment."[329] The motivation of substitute families must obviously be a prime consideration in order to protect children from abuse and exploitation. But the willingness of many families, even in impoverished circumstances, to take in an unaccompanied child, and their capacity to nurture the child has been recognised in many situations. In addition to the increased benefit to the children, such arrangements are less costly, and can direct resources to local communities, thereby strengthening their own capacities to care for their children.

Efforts at family reunification may also be influenced by attitudes within the military. They may not wish to admit to numbers of children in their forces, so they conceal the issue, or may consider that children can be better helped by remaining the responsibility of the military.[330] In the latter

[329] See, for example, Bonnerjea (1994) op. cit for a discussion of factors to be considered in the placement of children with substitute families. Also the experience of Rädda Barnen described by U. Blomqvist, *Social Work in Refugee Emergencies. Capacity Building and Social Mobilisation: The Rwanda Experience.* In: M. McCallin (Ed.) (1996) op.

[330] For example, the NPA in Uganda (see case study for Uganda in M. McCallin (1995) op. cit.), and Rwanda where a special school has been established for children associated with the Rwandan Patriotic Army in the course of the genocide of 1994, rather than effecting procedures for immediate family tracing and reunification. (Secondary source)

situation, evidence indicates that in many instances the army gathered up unaccompanied children, at least initially as a 'protective' measure. Here, child welfare issues, in particular the importance of family life for the children, may need to be introduced in a sensitive manner with the military in order to involve them in the children's reintegration. Insistence on the immediate removal of the children could be interpreted as in some way negating their concern for the children's welfare. But where the military has shown such concern, this can be built upon, and at least their knowledge of where children were found could assist in family tracing and reunification.

ECONOMIC CONSIDERATIONS

To a disproportionate extent, child soldiers come from poor and/or marginalised families. It is these communities which undoubtedly suffer the worst deprivations as a result of conflict, and face increased impoverishment as the conflict continues.

> *All parents are losing their sons, so families don't have enough labor to support themselves any more. Every family is getting poorer and poorer ... It's as though the SLORC is strangling the people to death. 55 or 60 boys already have to go out of every 100 households in my area.*[331]

Economic concerns can influence the process of demobilisation of children, even when families do not support the participation of their children. In Afghanistan at present, despite the fact that "people hate the war", there are no alternative sources of employment, and involvement in the conflict is a source of income.

The capacity for communities to accomplish economic recovery can have significant implications for those children who participated in the conflict. The need to contribute to the family's well-being can affect, for example, their ability to take advantage of even the limited opportunities for educational or vocational enhancement that are available.

> *Due to the heavy human loss in Afghanistan a large number of children*

[331] Burma/Myanmar case study

are now the earning members and have to work hard to support their families.[332]

Also, the time spent with the military may mean that the "child soldier" is actually a young adult when she/he returns home. The effect of long periods of time with the military may be that children and young people find themselves "alienated from the traditional means of livelihood available for them"[333] as they have never had the opportunity to learn to do agricultural work or obtain the skills which underpin the economic basis of their communities. Here there is an obvious link between economic considerations and reintegrating into family life and civil society. For most communities, "a concern for broader social and economic reconstruction (is) the most significant factor in the remaking of their worlds."[334] Allied to this is the extent to which children will be "valued for their contribution to the productive work of the family."[335] Thus where children can assist their families in economic recovery, the process of reintegration may run more smoothly. The children's capacity and willingness to work will be of immediate and appreciable value and will contribute to their re-attachment to the family and community, as well as being an "opportunity to 'deconstruct' the past", and to "develop an alternative meaning to their lives"[336] that is not associated with the power wielded by the gun.

EDUCATION AND VOCATIONAL TRAINING

In general, time spent with the military is seen as a time of lost opportunity. The theme of lack of skills learned in the military was pervasive, and examples of efforts to provide even limited education during conflict were few.[337]

I didn't receive any education apart from basic training ... I don't think

---

[332] Case study for Afghanistan.

[333] Case study for El Salvador

[334] S. Gibbs. *Post-War Social Reconstruction in Mozambique: Reframing Children's Experience of Trauma and Healing.* Disasters Vol. 18 No. 3 (1994)

[335] Gibbs op. cit.

[336] McCallin (1995)

[337] El Salvador, Burmese opposition forces, PKK

*there are any post-service educational opportunities for child soldiers,
because by the time you get out you're too old.*[338]

*Not through any fault of their own the adjustment they have to make is
difficult. Unemployment is high among former members of the self-
defence units. Their lack of education and training is particularly debili-
tating. The dislocation of the educational system does not help them in
any way and many teachers are not equipped to deal with the difficulties they experienced and exhibit.*[339]

This last reference touches on two issues that can have an impact on the
child soldier's rehabilitation and reintegration. Not only have the children
received no or limited education as soldiers, but in many countries the conflict destroys those institutions which serve children – in this case education – making it more difficult for them to resume normal life. Also, the
individuals whose role is to assist children, here the teachers, are themselves affected, and may find it difficult to make the effort to assist children with special needs. There is obviously here a requirement to target
the reconstruction of the educational system as a priority, not only for former child soldiers, but all children, and to ensure that teachers are given
support and, where necessary, additional training to cope with the added
demands of teaching children who have been affected by conflict.

*(The teachers) are aware that many have problems concentrating and
are accepting and understanding if a child becomes restless or needs to
leave class for a break. In conversation with the children, usually the
first thing they want to show a visitor is their school books. They see education as their way out of poverty and are thus motivated to learn ... The
activity of going to school ... is clearly having a major organising effect
on the children's behaviour and self-esteem ... It is important, therefore,
that the teachers are given the assistance they need to be able to teach.*[340]

[338] Burma/Myanmar case study
[339] Case study for South Africa
[340] Jareg, E. & McCallin, M. (1993) *The Rehabilitation of Former Child Soldiers.*
Report of a Training Workshop for Caregivers of Demobilised Child Soldiers.
Freetown, Sierra Leone. Geneva, International Catholic Child Bureau.

Given their limited educational opportunities, and the need in almost all situations to contribute to the family's economy, "school" for many former child combatants will likely require a vocational component to assist the children to acquire skills and find employment.[341] A number of the case studies made reference to the need to establish a secure economic base after conflict, and how this can influence the children's capacity to benefit from educational and vocational programmes.[342] To secure effective reintegration, and in some cases prevent re-recruitment or criminal behaviours, it will be very important to recognise the inextricable link between economic and educational concerns. Children miss out on education because of their participation, leaving them with few or no skills to secure employment and thus further marginalised. They are under pressure to contribute to the family's economy and, where participation in the conflict remains an option, will be unlikely to relinquish this source of income.

> *A large proportion of child soldiers in Afghanistan are illiterate. Since their present age is not appropriate to pursue education, they still prefer to be involved in the current conflict. It will take a long time to educate and encourage child soldiers to cease their involvement in the conflict. They need to be encouraged and provided with a special educational programme and other activities ... They should also be provided with alternate employment during the course of their studies to support their families.[343]*

Not wanting to return to school because there will be too much effort involved in catching up on lost time, or because they may be targets of abuse or victimization because of their involvement,[344] can affect the

---

[341] This issue is one which will be dependent on the age of the children, as many (at least in theory) will be too young to enter the employment market. See M. McCallin (1995) for a discussion of formal, non-formal and traditional apprenticeship training schemes as a means of combining vocational training with the children's need for basic education.

[342] See case study for El Salvador where only 6% and 32% of those eligible for basic and vocational education programmes respectively ectually enrolled.

[343] Case study for Afghanistan

[344] Uganda case study

extent to which former child solders take advantage of educational opportunities.

The importance of re-establishing educational systems must be noted because they provide an environment where children can gain "cultural literacy" and which "allows children to gain a sense of social responsibility".[345] Whilst these important aspects of education were noted in the context of on-going conflict, they are crucial components of effective rehabilitation and reintegration. Where communities are striving to normalise daily life "the organised structure of schools also helps to provide children with a sense of security essential to healthy child development".[346] It sends a positive message to children that their needs and welfare are matters of concern. Schools can also be the place where children develop "greater feelings of empowerment for strength and competence". They are also the natural setting for peace education. "Resolution" of a conflict does not imply that peace automatically breaks out, and violence may be a continuing characteristic of the children's lives.[347]

Restoring and, where necessary, restructuring educational opportunities for child soldiers will be a function of local conditions, and the ages, numbers and gender of the children recruited and the functions they performed. Whatever the process followed, there are alternative methods to traditional educational structures and systems that will need to be considered, including retraining of teachers to address special needs that arise due to participation in conflict.

CULTURAL AND SOCIAL ISSUES

The breakdown of traditional family and community structures as a result of conflict and in some cases the resulting empowerment of children, as happened for example during the Intifada and the struggle against apart-

---

[345] Case study for the Intifada

[346] Case study for the Intifada

[347] See, for example, the case study for South Africa where "Violence based on ethnicity was ... encouraged by the state" and there is now a need to reconcile "the different warring parties." Also, for a description of a programme to introduce peace education in primary schools, see: J. M. Tortorici. *Peace Education in Nicaragua.* In: M. McCallin (Ed.) (1996) op. cit.

heid in South Africa, can have profound implications for the children's rehabilitation and reintegration. Parents are rendered powerless to protect their children, often with the deliberate intention of undermining traditional values, and creating tension and discord within families. Post-conflict it appears unlikely that children who had engaged in activities beyond their years will readily submit to the old ways, and communities will need to be enabled to incorporate what may be lasting changes to their social and cultural systems.

> *Throughout the country the youth were exhorted to become young lions in order to shake off the yoke of their oppression. Black youths became accustomed to the notion of violently engaging in conflict with forces of the state ... Violence became a socially sanctioned means of attaining change within black urban populations.*[348]

In many conflicts, families and communities themselves encourage children from an early age to incorporate beliefs and standards that protect their cultural and ethnic identity. Once the conflict has ended, they may be reluctant to submit to restraints upon their behaviour, seeing themselves as the prime movers in effecting the desired social and political change.

> *The long-running civil war has meant that military service is something of a legacy, passed on from generation to generation ... Children who grow up in communities ... which face frequent attack and relocation develop political sensitivity against the central government at a very young age. They are raised to associate military induction with the survival of their ethnic culture.*[349]

Contrasting with such situations are those where children from certain ethnic or minority groups are targeted with the purpose of destroying their social structure. Here many young people may be permanently alienated from their communities of origin.

> *... forcing under 18s (from indigenous communities) to enrol in the army*

---

[348] S. Mokwena, *Marginalisation, Youth and Violence.* Quoted in case study for South Africa

[349] Burma/Myanmar case study

practically severs and destroys the future of these communities ... the process of acculturation, side-lining and dehumanisation that the youths are subjected to signifies a form of ethnocide. For these youths lose their identity, being ashamed of their origins and ancestral roots, often not returning to their communities on leaving military service.[350]

Even when this process is not deliberate, it must be recognised that for the children the separation from their families implied by their participation in conflict means that they lose "access to social and cultural values, which is even more serious given that (they) were in the crucial stages of development."[351]

Communities must be enabled to recognize the long-term consequences of the induction of their children into "... a political culture defined by the idea that established authority is illegitimate ..." in order that their children may realistically benefit from the opportunities available in the period of post-conflict resolution, and not be hindered because they "can't suddenly discover the virtues of co-operation merely because the nature of the authority has changed".[352]

The contrary situation to this is one where children are considered difficult because they lack initiative "having been under military discipline for a long time, and having been provided with all basic necessities".[353]

During military service, the children are separated from the supportive and nurturing environment of the family, and are dependent on rigid and authoritarian structures to control their behaviour ... This is the way of life they know, and the one on which they model their own behaviour towards others. Also they were unused to making decisions for themselves and may initially exhibit a continued dependency on authoritarian structures to control and limit their behaviour. Once this is removed, and without a support structure in place, the risk may be that not having learned accepted mechanisms for personal control they may resort to aggressive techniques to get what they want.[354]

[350] Case study for Guatemala
[351] See case study 2 for Mozambique in M. McCallin (1995) op. cit.
[352] Case study for the Intifada
[353] Case study for El Salvador
[354] M. McCallin (1995) op. cit.

A number of the case studies made reference to negative attitudes within their communities as a result of the children's difficult, violent or even criminal behaviour. Expectations of a smooth and trouble-free progress towards social reintegration will in many cases be unrealistic. The cumulative effect of their experiences as soldiers, their separation from their families, and the likelihood of inadequate provision of basic services upon return will mean that, for many, social reintegration is the most difficult phase they encounter in a process fraught with risks to their well-being.

> *More (disturbing) ... is the fact that they are being seen as the persons responsible for the crime wave. This is not so says a former unit member ... There are some members but then their behaviour stems from other factors and not merely because they are members of the SDUs. One of these is the fact that many former members ... find it difficult to return to school.*[355]

One of the most difficult issues that some communities will confront when children return home is whether and how children are held responsible for the acts they committed as soldiers. Recent work by Save the Children (USA) in Rwanda indicates that "the application of *legal justice* might not serve the requirements of the *social justice* expected at the popular level".[356] Whilst this work may be considered to describe an extreme case, reflecting conceptions of the culpability of children who participated in acts of genocide, it has applicability particularly in those situations where there is an ethnic or tribal dimension to the conflict, or where formerly opposed factions are now required once again to live side by side. Because we know so little about what happens to children after demobilisation, the issue of juvenile justice as a factor influencing social reintegration was not documented in the research. It requires further study to ensure the protection of child combatants following demobilisation, and to promote post-conflict reconciliation between opposing groups.

[355] Case study for South Africa
[356] *Children, Genocide and Justice. Rwandan Perspectives on Culpability and Punishment for Children Convicted of Crimes Associated with Genocide.* Save the Children Federation (USA). Kigali, 1996.

They tend to be aloof from others with little confidence in themselves and distrustful of others. They tend to lapse into absentmindedness as well as swift mood changes for no apparent reason ... Some ... are reported to be undisciplined and difficult to control by parents or guardians. One such group of boys has acquired guns and is laying ambushes on passers-by ...[357]

There is a huge difference between the educational level, skills, behaviour and way of thinking of a child soldier and a child not directly involved in armed conflicts.[358]

The physical and emotional well-being of child soldiers at the time of demobilisation and reintegration is influenced not only by the experiences inherent in direct participation in conflict, but also the manner in which they as individuals were treated by the factions which recruited them. The impact of "...physical and emotional abuse... used to reiterate the importance of rank and subordination, involving debasement and humiliation of the younger recruits",[359] can result in low self-esteem, guilt feelings and violent solutions to problems.[360] Lack of trust and confidence in others, and aggressive or militaristic behaviour may be the real or supposed consequences of such experiences. The latter is often the preferred, or at least presumed, effect of children's direct participation in conflict, as evidenced by attitudes of people involved in the rehabilitation of former child soldiers in Nicaragua:

Until that moment we only knew what was said about them: "bloodthirsty children", "humantigers", "they take out people's eyes", "born assassins" ... According to the first director of the project, the personnel had even been warned of the need to prepare themselves physically by taking karate or self-defence lessons so as to be prepared for any possible aggression.

[357] Case study for Uganda
[358] Case study for Afghanistan
[359] Burma/Myanmar case study
[360] Uganda case study

An intoxicated young rebel with the United Liberation Movement of Liberia (ULIMO) in 1996. In many wars, child soldiers are supplied with drugs routinely or before battles. Photo: Reportagebild.

Reintegration is seen as less of a challenge where socialisation within an opposition movement has given rise to a sense of empowerment, and where "the movement too becomes family to the children".

> ... child soldiers are treated with respect and as human beings ... benefits are equal to or translated to mean self-improvement, self-esteem and confidence, respect, the opportunity to serve the masses, the poor.[361]

In this regard, the meaning of the events which the children have experienced, and their understanding of these events, can act as a buffer to reduce negative consequences for their emotional well-being. One theme which did emerge was particular to certain rebel or opposition forces who did attempt to give some meaning to the children's participation – an awareness of fighting for democracy, a sense of shared purpose and belonging – "the relationship and social welfare that I ... experienced ... will be useful".[362]

> ... an ideological explanation of violent conflict, along with a sense of being on the "right side", helped to buffer Palestinian children against some of the negative psychological effects of violence. Such a cognitive framework was important in helping children work through their reactions to violence and gain a sense of competence and mastery.[363]

In most cases the physical well-being of child soldiers is seriously compromised, due to inadequate nutrition and health care, and the effect of harsh punishments[364] and training regimes. Children suffer from worm infestation, respiratory tract infections and water-borne diseases due to poor hygienic conditions and lack of basic facilities. Many are known to be suffering from sexually transmitted diseases, and it may be supposed that this

---

[361] Case study for the Philippines

[362] Burma/Myanmar case study

[363] a) Baker and b) Anthony & Cohler in case study for the Intifada

[364] In Liberia a form of torture known as 'tabay' was practised on children. This involves tying the elbows together behind the back for prolonged periods. The result is nerve damage to the arms. Cited in: *Easy Prey. Childs Soldiers in Liberia.* Human Rights Watch/Africa & Human Rights Watch Children's Rights Project. (1994)

problem will be common to almost all cases due to the prevalence of sexual abuse of child combatants reported in the case studies. The routine administration of drugs and alcohol to enhance the "performance" of child soldiers in combat[365] will also have an impact on their physical and psychological well-being. The long-term implications of substance abuse, where "... young soldiers were given amphetamines, tranquilizers and alcohol before being sent into battle"[366] are considerable, particularly when the availability of such substances after demobilisation may compromise the process of rehabilitation and reintegration.

Of particular concern when one considers social reintegration will be the extent of physical disabilities that child soldiers suffer. Previous sections have described the appalling injuries inflicted on children. Their disabilities will in fact pose additional problems for them, and may result in their marginalisation or rejection by their families and communities who are unable to cope with the additional burden of a disabled family member.

> N was 14 when he became a child soldier. Having lost his hearing and two hands in combat, married with two children, he finds himself ... with no work and no resources. "Nobody will give work to a former militiaman, we are avoided like the plague. Yesterday I had to stop sending my children to school. I did not have any money to pay for their transport." [367]

Sexual abuse of the children which, in some case studies, is reported to be widespread, will have a particularly serious impact on the capacity of girl soldiers to reintegrate with their communities, as their families may be unable to accept or acknowledge their use as "wives" or "comforters" of the male combatants.

> Girl children are more vulnerable to sexual abuse than adult women or maturer girls who are shunned as having been already "used" by men and who probably could have HIV/AIDS already ... In some instances

[365] Liberia
[366] Burma/Myanmar case study
[367] Case study for the Lebanon

15-year-old boys have been allocated girls as wives. Such boys become proud and happy though they are likely to end up with sexually trans-mitted infections.[368]

... there were girls who were obliged to have sexual relations with the combatants ... The women among other things had to "alleviate the sadness of the combatants." And who alleviated our sadness after going with someone we hardly knew? ... At my young age I experienced abortion. It was not my decision, I could not decide on that. They decided, in any case. Hadn't I handed over my entire life? Had I not undertaken a commitment to permanent obedience and discipline? ... There is a great pain in my being when I recall all these things, principally because with time I have come to understand that to be a woman in any group was always a disadvantage. In spite of my commitment, they abused me, they trampled my human dignity. And above all, they did not understand that I was a child and that I had rights.[369]

Cultural beliefs and attitudes may affect the girls' opportunities for marriage, with the result that they do not stay with their families and see prostitution as their only means of livelihood. Additionally, communities may find it difficult to accept that their daughters were also combatants. Many young women, whose identity was forged in combat, may have to conceal the role they played in conflict and, "for fear of total rejection by their husband's family (must) pretend to be the gentle, soft-spoken and submissive woman that their civilian counterpart is."[370]

The issue of long-term impacts of sexual abuse of the boys is one about which little is known, and which will require further sensitive enquiry, in order to take account of cultural mechanisms to deal with such a delicate problem.

[368] Case study for Uganda

[369] Former female combatant, interviewed for the case study for Honduras

[370] P. Maramba, (1995) *Reintegration of Demobilized Combatants into Social and Economic Life. The Zimbabwean Experience.* Report on a consultancy for the ILO, Geneva.

It will be apparent that the social reintegration of child soldiers into families and communities devastated by the effects of conflict will be a far from straightforward process. In particular there is a danger that a focus on their "traumatic" experiences, and the psychological consequences of these experiences, may lead to the implementation of a Western therapeutic or treatment model to address their situation. As with other groups of children made vulnerable because of armed conflict, there is a number of considerations that affect such an approach. The arguments against such a model focus on its cultural inappropriateness, and the resulting marginalisation or even "misdiagnosis" of many children. Other concerns point to the fact that it does not incorporate the knowledge and expertise of the local community, is costly to implement, reaches only a small proportion of the population, and contributes little to building capacities in local communities.

The alternative strategy, whilst recognising the impact of traumatic events, considers that children's responses may be influenced as much, if not more so, by their experiences of family loss and separation, and the precarious nature of family life that results from conflict.[371] Thus approaches to healing and recovery focus on re-establishing nurturance and support from within the child's immediate environment of home, family and community.[372] There is now considerable experience available on community development approaches with war-affected and displaced people which indicates that they provide a more flexible response that takes account of the social context, and incorporates the community in the process of defining and implementing solutions to their situation.

[371] For a discussion of "secondary stress factors" that children may experience see: E. Jareg, & P. Jareg, (1994) *Reaching Children through Dialogue*. McMillan Press.

[372] Significant expertise has been developed in recent years within international organisations such as UNICEF, and local and international NGOs on the implementation of community-based approaches which is relevant to the social reintegration of child soldiers. See, for example, the bibliography in the Save the Children Alliance paper mentioned above. Also, S. Fozzard, (1995). *Surviving Violence. A Recovery Programme for Children and Families.* Geneva, International Catholic Child Bureau.

Typical characteristics of community development approaches are:

- an emphasis on the need for *long-term development* and not just the more immediate relief of suffering;

- a strong emphasis on the importance of the *social context* in which people experience stress, and not just individual suffering;

- seeking *community definitions of needs and priorities* rather than professional judgements about them;

- seeing *traumatic experience as one among many aspects of stress* faced by people: distressing experiences of violence may be compounded by current difficulties such as poverty, continuing fear, poor housing, unemployment, lack of social resources;

- emphasis is placed both on the *collective nature of problems* and the need to promote *collective coping mechanisms.* It sees communities as resourceful and potentially powerful;

- community-oriented programmes tend to focus on *common needs within the community* rather than the specific needs of individuals.[373]

Over and above theoretical considerations of the relative value of these contrasting approaches, generally conducted in the sanitised environment of conferences in North America and Western Europe, it is the near-universal acceptance of the Convention on the Rights of the Child which, in effect, defines the applicability of community-based interventions. This is most evident in the "best interests" provisions of the CRC.

If, according to the best interests of the child, children and their families are positioned at the centre of all strategies to close the gap between reality and the optimum environment for full development, then a model

[373] Reproduced from: D. Tolfree, (1996). *Restoring Playfulness. Different Approaches to Assisting Children who are Psychologically Affected by War or Displacement.* Rädda Barnen, Stockholm.

that treats symptoms rather than empowers whole human beings may be
inadequate and counterproductive ... (T)he medical model ... tends to
universalize a Western notion of child development which, even by
using the term trauma, pathologizes children's invisible wounds and
views them only as passive victims rather than active survivors.[374]

To respond to the needs of the children from the limited perspective of a
trauma or sickness model is to deny the complexity of their situation, and
a failure to recognise that the abuse of children as soldiers is, in almost all
instances, a consequence of poverty, repression and social injustice. More
than psychotherapeutic interventions, their situation and that of their families cries out for interventions that improve their conditions of life, and
acknowledge the role of the family and community in effecting the recovery of children who have experienced distressing events.

Within such a framework, international organisations, rather than co-
opting to themselves the role of the principal actors in developing programmes of assistance and intervention, can focus on facilitating attention
to the needs of the children through dialogue, partnership and advocacy
with affected communities. In this way the "spirit" of the Convention on
the Rights of the Child is translated into practical realities that promote
the well-being of the children. Rather than a remote instrument of international humanitarian law, its implementation at local and national levels,
both as a tool in advocacy and as the reference point for guiding assessment
of the situation of children, ensures that children benefit from adherence
to its provisions.

The above issues principally describe child welfare concerns that influence the social reintegration of child soldiers. It is important to recognise,
however, that they cannot be addressed simply from a child welfare perspective implemented on an *ad hoc* basis in each situation. They require
more determined advocacy to implement the overall provisions of the

---

[374] For a full discussion of the relevance of the CRC to this issue see: D.
Reichenberg, & S. Friedman, *Traumatized Children. Healing the Invisible
Wounds of Children in War: A Rights Apporach.* In: Y. Danieli, N. S. Rodley, & L.
Weisaeth, (Eds.) International Responses to Traumatic Stress. Amityville, New
York, Baywood Publishing Company, Inc. 1996

Convention on the Rights of the Child in order to ensure "the protection and harmonious development"[375] of the children. The issues are complex and interrelated, and require a political will at both local and international level to address the conditions of life that not only lead to recruitment, but also compromise the children's rehabilitation and social reintegration. We need to create a more informed understanding not only of the social and political conditions that give rise to recruitment, but also of the consequences for children's participation in conflict, in order to facilitate attention to their needs and protect them from further harm.

One aspect of protection in this context is to raise awareness that there are ethical issues to be considered in achieving a balance between directing attention to the plight of the children, without exploiting any individual child. All too often, interest in child soldiers focuses on the sensationalistic nature of their involvement in the conflict – the acts they committed – with little explanation of the circumstances that lead to recruitment or the consequences of participation for the children's well-being.

> We often come across children who have been pressured into telling and re-telling their "horror stories" to journalists, researchers and sometimes even to officials of aid agencies and psychologists ... Field staff cannot avoid the need to collaborate with researchers and journalists: it is the terms of such collaboration that require clarification.[376]

This "retelling" of his story is not good child welfare practice. In the course of a visit by one of the authors to the school in Rwanda that has been established for children associated with the RPA, the Director expressed concern about the number of visits to the school by journalists and representatives of foreign organisations. Children had been photographed and questioned about their experiences. They were often left distressed as a result, and did not understand why there was no follow-up to these visits. Repeated questioning about events in which he participated may compro-

---

[375] Preamble to the UN Convention on the Rights of the Child

[376] op. cit. Paper prepared by Save the Children Alliance (1996). This paper also presents guidelines on interviewing children.

mise a child's rehabilitation and reintegration as he sees that he is valued only with regard to his identity as a soldier, or worse, a "killer", and at times may lead to children "embellishing" what are already horrific events to gain attention. The child's physical security may also be put at risk, if he can be readily identified from photographs or other details about his background. Such situations infringe the child's right to privacy, and his right for recovery to take place "in an environment which fosters the health, self-respect and dignity of the child."[377]

[377] Articles 16 & 39 of the Convention on the Rights of the Child

Islamic Jihad holds a memorial for a Palestinian killed in a suicide bomb attack. Suicide missions are the particular province of adolescents.
Photo: Gamma/IBL.

CHAPTER 6

Military Attitudes to Child Soldiers

A DIFFERENCE BETWEEN FORCED recruits and volunteers which is often merely technical but sometimes can be quite significant is that the volunteers are not picked out by the military but present themselves for recruitment either spontaneously or as a result of active canvassing for recruits. In either case there is the opportunity for the military to vet those who present themselves, with a greater or lesser degree of selectivity.

In some instances, there is no evidence of volunteers being turned away as too young, or for any other reason. In others there is at least a nominal adherence to minimal recruiting ages, and some specific instances of refusal or diversion are given in the section on preventive strategies below.

Various reasons are given for the positive choice to recruit children, either forcibly or as volunteers:

... children are much more easily moulded than adults, who are already "corrupted by the ideas of capitalism".[378]

They can easily be trained and disciplined, have a high degree of camouflagibility, and can run risky missions. The most significant factor however is [that they are] easy to mobilise whether recruited forcibly or voluntarily.[379]

... children are easier to intimidate and scare with punishments and

---

[378] Case study for Peru (translated from the Spanish)
[379] Case study for for Ethiopia

they obey more easily. Also, children are less inclined to desert than adults. As children have few opportunities outside the barracks, especially in relation to work, it is easier for them to become adapted to their new life conditions, rather than an adult recruit taken from his work, studies or family.[380]

... the young fighters normally have no family responsibilities and are healthy. They are also easily controlled and they naturally obey adults.[381]

Some ... military commanders claim that child soldiers are more obedient, do not question orders, and are easier to manipulate than adult soldiers.[382]

Children are an attractive option for guerilla leaders as they are easily conditioned and motivated, and can be put into battle with less training. Although the children are put into more danger, they too are of greater danger to their adversaries.[383]

In some contexts, as previously discussed, children, especially the youngest recruits, are seen as having a particular value which mature soldiers do not possess, as spies, decoys, lookouts, minesweepers, and so on, while teenagers are recruited and trained specifically for assassinations and suicide attacks.[384]

There are, however, opposing viewpoints. Not all commanders share the assessment that children can be tractable troops with particular advantages in certain circumstances. For some, their employment is morally repugnant. For others, they are seen even in purely military terms as an encumbrance. It is noteworthy that the only case study to include the results of widespread interviews with (opposition) commanders revealed a number of dissenting opinions:

[380] Case study for Paraguay (translated from the Spanish)
[381] Case study for Liberia
[382] Case study for Burma/Myanmar
[383] Case study for Sri Lanka
[384] Case studies for Lebanon and Peru

154

I try as much as possible to avoid having the really young ones in my unit. They just can't carry the heavy artillery. Kids slow us down. We like to keep them in the kitchen, or doing chores back at base.

[Some commanders] state that child soldiers are still children, and that they often misbehave, are unruly, and require frequent scolding, as young boys usually do. One ... Captain claimed that child soldiers often had to be taught the most basic life skills, such as how to wash, cook and take care of themselves.

Children are mostly good fighters, but they are not always careful ... When there is shelling, the younger ones forget to take cover. They get too excited. They have to be ordered to get down inside the bunkers.

Both the lifestyle and the duties of military service are extremely demanding and, with a limited force of soldiers and arms, ability to perform is critically important.

Some really young children ... are not ready or suitable for this kind of work. I send the ones who are not ready to school here, and the army funds their schooling. As for the children who do not seem cut out for this work, I arrange for one of the older boys to plan a "desertion" or escape as they will not leave without a leader.

In my regiment, there are some 12 and 13 year olds. I try to send them to school if I can; then they can join and become a soldier later if they decide to, or return to their families.[385]

These are not the only armed opposition groups opposing the involvement of children:

[The] policy of the liberation movement was actively to discourage youth, and in many instances including those over 18, from joining [their armed wings]. A direct result of this policy ... was the establishment of the ... College of Education in 1978 to educate the youth ... Persons under the age of 18 years who wished to form part of the armed

---

[385] Case study for Burma/Myanmar

wing were actively counselled against joining the armed wing and rather encouraged to continue their schooling or studies.[386]

Furthermore, the attitude of commanders may change with experience of the difficulties of controlling children with access to arms and bomb-making equipment, and the greater likelihood of them succumbing if caught:

The refinements of [government] interrogation techniques was another factor which forced [the armed opposition group] to rethink on the role of the [youth wing]. Youths appeared more susceptible to interrogation. The result was that in 1976 numbers in prison rose dramatically; a high percentage of these were youths.[387]

As children they are prone to adventurism and abuses. They tend to abuse the use of guns. The young rebel surrenderees, the ones who benefitted from a weak political foundation, are now into illegal gambling.[388]

Some professional soldiers share these reservations about the participation of children:

Leaders of the coup [of high-ranking officers] stated their reasons for the coup attempt as mainly being the stalemate in the armed conflict and the increased involvement in the armed forces of children whom they see as immature and negatively bearing on the morale and effectiveness of the armed forces.

... members of the regular force are often heard attributing the defeats and unruliness of the [government] army to the fact of recruitment and service of children ...[389]

[386] Case study for South Africa
[387] Case study for Northern Ireland
[388] Case study for the Philippines
[389] Both from the case study for Ethiopia

CHAPTER 7

Legal Standards on Recruitment and Participation

THERE ARE TWO TIERS of applicable law: national law and international law. Most national law on recruitment is clear and straightforward, and predominantly sets 18 years as the minimum age for recruitment, although there are a few exceptions with lower or higher minimum ages and somewhat more who permit volunteers below the age of 18.[390] Even in these cases, some governments have a policy of not sending under-18s into combat.[391]

However, there are too many instances in which the law is ineffective. One problem is that in the world today, armed conflict is almost exclusively in the form of internal rather than international armed conflicts. This highlights the practical limitations of national legislation on recruitment. When a government regulates recruitment into armed forces, it is only addressing recruitment into its own armed forces. This means that recruitment into armed opposition groups (which are, by definition, illegal in any case), is not covered by the law, nor are any military regulations penalising illegal recruitment applicable to those recruiting for such groups. When legislating, governments do not normally contemplate the possibility of civil war (other than by creating offences such as treason) and, therefore, do not

---

[390] See Annex 1
[391] For example, Australia and the Netherlands

attempt to regulate recruitment by armed opposition groups, for example through the criminal law.

There are two branches of international law which are relevant to the issue of recruitment ("recruitment" covers both compulsory and voluntary enrolment)[392] of children into armed forces: international humanitarian law – the law which governs the conduct of armed conflicts – and human rights law. At present, under international law the recruitment of children into armed forces and armed opposition groups and their participation in hostilities ("hostilities" means a battle or similar armed engagement in an armed conflict) is governed by three legal instruments: Additional Protocols I and II to the Geneva Conventions of 1949, and the Convention on the Rights of the Child.

The Convention on the Rights of the Child was adopted by the UN General Assembly in 1989 and nearly every country in the world is now a party to it.[393] It is legally binding on every government which is a party to it and applies to all children within the jurisdiction of each state party, not only to those which are nationals of that state. Article 38 of the Convention provides, in part:

(1) *States Parties undertake to respect and to ensure respect for rules of international humanitarian law applicable to them in armed conflicts which are relevant to the child.*

(2) *States Parties shall take all feasible measures to ensure that persons who have not attained the age of 15 years do not take a direct part in hostilities.*

(3) *States Parties shall refrain from recruiting any person who has not attained the age of 15 years into their armed forces. In recruiting among those persons who have attained the age of 15 years but who have not attained the age of 18 years, States Parties shall endeavour to give priority to the oldest.*

---

[392] See for example, M. T. Dutli, "Captured Child Combatants", International Review of the Red Cross, Sept–Oct 1990, pp 421–434, p 424

[393] The exceptions are Somalia and the USA

Paragraph (1) derives from Article 1 common to the four Geneva Conventions of 1949 and paragraphs (2) and (3) from Article 77(2) of Additional Protocol I to those Conventions. Therefore, the existing jurisprudence and interpretation of the Geneva Conventions and Protocols applies to Article 38 where relevant.

Article 38 was an innovation in explicitly incorporating international humanitarian law into international human rights law. Previously it was necessary to draw on general provisions requiring states parties to human rights treaties to abide by their other obligations under international law in order to make the links.[394]

Article 38(1) applies the dual standard of requiring the states parties both to respect and to ensure respect for these provisions. It therefore establishes positive as well as negative duties both with regard to their own conduct and also in relation to the conduct of others (whether other governments or armed opposition groups). The limitation to "conflicts relevant to the child" is superfluous. It is hard to conceptualise an armed conflict which is not "relevant to the child", even apart from the difficulty of defining "relevant". Leaving aside the situations in which children are directly affected by attacks, recruitment and so on, armed conflicts tend to have severe impacts on the population in general, on the provision of services, including food, health-care and education, and on infrastructure, and also tend to affect adults who are relevant to the child, such as fathers and brothers. The logical interpretation of Article 38(1) is that it simply reinforces the obligations of states to abide by the international humanitarian law by which they are already bound. As such its value is in making the explicit link between these two branches of international law affecting the human person and in creating the possibility that the Committee on the Rights of the Child, the supervisory body established by the Convention, could examine states' compliance with their obligations under internation-

---

[394] For example, Article 4 of the International Covenant on Civil and Political Rights which permits states to derogate from some of their obligations "In time of public emergency threatening the life of the nation ... provided that such measures are not inconsistent with their other obligations under international law..."

al humanitarian law, at least in so far as relevant to the child. (This is where the limitation does makes sense).

No definition or delimitation of the term "armed conflict" is provided in the Convention on the Rights of the Child. It can be argued that a factual definition[395] therefore applies rather than the technical legal definitions of international humanitarian law (see below). The Convention therefore avoids the problem of governments denying that they are involved in an armed conflict as classified by the 1949 Geneva Conventions or the 1977 Additional Protocols, and therefore denying the applicability of international humanitarian law. One of the consequences of such denial is that it is harder to hold armed opposition groups to any rules since strictly speaking international human rights law is only binding on governments and why should the opposition accept the application of international humanitarian law if the government does not.

In the world today, armed conflict is predominantly in the form of internal rather than international armed conflicts and many of the child soldiers are serving in armed opposition groups. Article 38 does not appear to govern recruitment by armed opposition groups since Article 38(3) only addresses states parties' recruitment into their own armed forces. However, Article 38(2), particularly when read in conjunction with Article 38(1), does require states parties to the Convention to take all feasible measures to ensure that under-15s do not take a direct part in hostilities, without any apparent limitation to their own armed forces, nor to the type of armed conflict in which the state is engaged since Article 38 makes no distinction between international and non-international (internal) armed conflicts. The boundaries of what "feasible measures" and "direct" participation constitute are, of course, open to debate. However this means that a state party to the Convention has an obligation to try to prevent the direct participation of under-15s in hostilities on its own side, in armed opposition groups it is fighting, on the side of any other government or of any other armed opposition group. In other words the obligation is not only on the govern-

---

[395] That is, by assessing the actual situation on the ground rather than judging whether certain legal conditions are fulfilled or whether the government accepts that it is involved in an armed conflict.

ment engaged in an armed conflict but on all those who are parties to the Convention in relation to anyone engaged in an armed conflict.

Article 38(3) is an absolute obligation on the governments not to recruit under-15s into their own armed forces. This should cover any organised armed group which is under the control of the government or which could be under governmental control and not only the regular armed forces. In particular, it should cover any paramilitary forces, local militias, civil defence committees, and so on, under whatever name, which the government encourages or allows, that is any armed group which the government establishes, condones, arms or permits to bear arms: the crucial question is simply whether the group is allowed to carry arms. Any group which meets these criteria can and should have its recruitment regulated by the government in conformity with Article 38.

The requirement in Article 38(3) to "endeavour to give priority to the oldest" when recruiting in the 15-18 age group is a weak obligation. At best, it could be said to require a system which takes account of age in recruiting in this bracket. Its greater use is probably for the purposes of advocacy and moral pressure.

A limitation of Article 38 of the Convention is that it is only binding on States Parties, that is on governments and not on armed opposition groups. However, the obligation in Article 38(1) incorporates, *inter alia*, Additional Protocol II for those States which are parties to both documents. Furthermore, Article 41 of the Convention provides: "Nothing in the present Convention shall affect any provisions which are more conducive to the realisation of the rights of the child and which may be contained in: (a) The law of a State Party; or (b) International law in force for that State". In other words, if greater protection to children would be given by some other legal instrument to which the state is a party, its provisions will take precedence over those of the Convention on the Rights of the Child. In addition, if the national law sets 18 years as the minimum age for recruitment, the government will be bound by that rather than the lower limit in the Convention.

Since the wording of Article 38 of the Convention on the Rights of the Child was taken from Additional Protocol I, Article 77(2) and that effectively all countries in the world are now parties to the Convention, it is not necessary to examine Protocol I separately as in effect this aspect of the Protocol has been superseded by the Convention.

Article 4(3)(c) of Additional Protocol II of 1977 states:

> *children who have not attained the age of 15 years shall neither be recruited in the armed forces or groups nor allowed to take part in hostilities.*

This provides an unequivocal and total prohibition on recruitment of under-15s into any type of armed forces (governmental or opposition) and any kind of participation by them in hostilities. Thus it does not limit the governmental obligation to taking "feasible" measures to prevent them taking "a direct" part in hostilities, nor does it limit the state's obligation to not recruiting under-15s into its own armed forces only. In addition, the obligation is also imposed on any opposing armed groups.

However, Additional Protocol II[396] only applies to internal armed conflicts in the territory of a state party to Protocol II between governmental armed forces and "dissident armed forces or other organised armed groups which, under responsible command, exercise such control over a part of its territory as to enable them to carry out sustained and concerted military operations and to implement this Protocol" (Article 1(1)). It does not "apply to situations of internal disturbances and tensions, such as riots, isolated and sporadic acts of violence and other acts of a similar nature, as not being armed conflicts" (Article 1(2)). The effectiveness of Protocol II is limited because of the high threshold for its application, the reluctance of governments to accept that they are involved in a civil war, and the fact that it only applies if the state is a party to the Protocol.

Finally, in relation to all these provisions it is important to stress that it is not the children who are recruited (even if they volunteer) or who participate in hostilities who are in breach of the provision but the government or other authorities who recruit them or permit their recruitment or participation.

---

[396] See Annex 1 for the current list of states parties to Additional Protocol II

At the time of drafting the Convention, the question of the minimum age for recruitment into armed forces and participation in hostilities gave rise to considerable controversy. Many wished to see the minimum age set at 18 years in line with the general age of majority stated in Article 1 of the Convention.[397] In fact, this is the only provision in the Convention which specifies an age lower than 18 years as applicable. Even in this provision, some recognition is given to the 18-year threshold for majority in so far as Article 38(3) requires states "to endeavour to give priority to the oldest" when recruiting in the 15–18 year age range. Since then, a number of steps have been taken to raise the minimum age to 18.

The only regional human rights treaty which addresses the age of recruitment is the 1990 African Charter on the Rights and Welfare of the Child, Article 2 of which defines a child as every human being below 18 years. Article 22 reaffirms the obligation to respect and ensure respect for rules of international humanitarian law, and goes on "(2) States Parties to the present Charter shall take all necessary measures to ensure that no child shall take a direct part in hostilities and refrain in particular, from recruiting any child". Thus governments may not recruit under-18s, and have an absolute duty to ensure that under-18s do not take a direct part in hostilities. Unfortunately, not enough states have yet become parties to this Charter to bring it into force.

In December 1995, the Council of Delegates of the Red Cross and Red Crescent Movement adopted by consensus a "Plan of Action concerning Children in Armed Conflict", which includes the Commitment:

To promote the principle of non-recruitment and non-participation in armed conflict of children under the age of 18 years.

This commitment is broken down into three specific objectives:

---

[397] Eight countries have expressed their wish for a higher age by way of declarations on signature or ratification – Argentina, Austria, Colombia, Germany, the Netherlands, Spain, Switzerland and Uruguay.

> *1. Promote national and international legal standards (such as
> an Optional Protocol to the Convention on the Rights of the
> Child) prohibiting the military recruitment and use in hos-
> tilities of persons younger than 18 years of age, and also the
> recognition and enforcement of such standards by all armed
> groups (governmental and non-governmental).*
>
> *2. Prevent children from joining armed forces or groups by offer-
> ing them alternatives to enlistment.*
>
> *3. Raise awareness in society of the need not to allow children to
> join armed forces or groups.*

The work involved in preparing and adopting this Plan of Action also led to the adoption (by consensus) of a Resolution at the 26th International Conference of the Red Cross and Red Crescent (Geneva, 3–7 December 1995), which in addition to supporting the drafting of an optional protocol to the Convention on the Rights of the Child "to increase the protection of children involved in armed conflicts",

> *c) also strongly condemns recruitment and conscription of chil-
> dren under the age of 15 years in the armed forces or armed
> groups, which constitute a violation of international humani-
> tarian law, and demands that those responsible for such acts
> are brought to justice and punished;*
>
> *d) recommends that parties to conflict refrain from arming chil-
> dren under the age of 18 years and take every feasible step
> to ensure that children under the age of 18 years do not take
> part in hostilities;*

Although such resolutions are not legally binding, they are significant because they are adopted at a meeting open to all states parties to the 1949 Geneva Conventions and 1977 Additional Protocols and representatives of all recognised national societies of the Red Cross and Red Crescent Movement.

The UN Study on the Impact of Armed Conflict on Children, mandat-ed by General Assembly resolution 48/157 of 20 December 1993, was pre-sented to the General Assembly on 8 November 1996 by Graça Machel, the

Young revolutionary guard during a demonstration in Iran. Photo: Abbas/Magnum.

expert appointed by the UN Secretary-General to undertake the Study. Specifically on child soldiers, the Study recommends *inter alia* "States should ensure the early and successful conclusion of the drafting of the optional protocol to the Convention on the Rights of the Child on involvement of children in armed conflicts, raising the age of recruitment and participation in the armed forces to 18 years (Para. 62(d)).

The optional protocol to the Convention on the Rights of the Child referred to is being drafted by a working group of the UN Commission on Human Rights. The fourth session of the working group was held in February 1998 but was unable to reach agreement. Although there was near consensus on raising to 18 years the minimum age for participation in hostilities, for compulsory recruitment (conscription) into government armed forces, and for any kind of recruitment into non-governmental (opposition) armed groups, this was blocked by the USA's refusal to accept 18 for participation. The questions of the minimum age for voluntary recruitment into government armed forces, whether there should be a further exception for recruitment into military schools and whether the prohibition on participation should be limited to taking a direct part in hostilities remain unresolved. As a result, the Chair of the Working Group has been mandated to undertake informal consultations to see whether the problems can be resolved.

Furthermore, the particular vulnerability of displaced children to recruitment (as evidenced elsewhere in this book) has been recognised in the "Guiding Principles on Internal Displacement", developed by the Representative of the UN Secretary-General on Internally Displaced Persons. Principle 13(1) states: "In no circumstances shall displaced children be recruited nor be required or permitted to take part in hostilities."

Finally, the Statute for an International Criminal Court was adopted in Rome in June 1998. It will enter into force when 60 States have become parties to it. The creation of this Court could be a breakthrough in preventing the recruitment and use in hostilities of under-15s since the Court will be able to try individuals for committing, ordering, soliciting or inducing the commission of the crimes within its jurisdiction, or aiding, abetting or assisting in their commission. The category of war crimes in international armed conflicts includes:

Art. 8(2)(b)(xxvi): Conscripting or enlisting children under the age of fifteen years into the national armed forces or using them to participate actively in hostilities.

The equivalent provision for non-international armed conflicts is:

Art. 8(2)(e)(vii): Conscripting or enlisting children under the age of fifteen years into armed forces or groups or using them to participate actively in hostilities.

Non-international armed conflicts are defined for this purpose as "armed conflicts that take place in the territory of a State when there is protracted armed conflict between governmental authorities and organised armed groups or between such groups" and "thus does not apply to situations of internal disturbances and tensions, such as riots, isolated and sporadic acts of violence or other acts of a similar nature", Art. 8(2)(f).

WHY 18?

The reason for seeking to establish 18 as the minimum age for recruitment and participation in hostilities is not only based on the general age of majority stated in Article 1 of the Convention on the Rights of the Child. In relation to recruitment, Article 38(3) marks out the 15–18 year age group and exhorts states recruiting in that age bracket to give priority to the eldest. The age of 18 is already recognised under international law as being a significant demarcation line. A number of provisions prohibit the application of the death penalty to those under the age of 18 at the time the offence was committed.[398] ILO Convention No. 138 on Minimum Age, 1973, sets 18 years as "the minimum age for admission to employment or work which by its nature or the circumstances in which it is carried out is likely to jeopardise the health, safety or morals of young persons". Although the armed forces are considered to be outside the legal scope of this Convention, the ILO itself has suggested that it "may be applied in corollary to the involve-

[398] Convention on the Rights of the Child, Article 37(a), International Covenant on Civil and Political Rights Article 6(5), Additional Protocol I Article 77(5) and Additional Protocol II Article 6(4)

ment in armed conflicts".[399] Similarly, in Article 7(2) of the revised European Social Charter, states parties undertake to "provide that the minimum age of admission to employment shall be 18 years with respect to prescribed occupations regarded as dangerous or unhealthy", and the Council of Europe has stated, "In view of the specific nature of military activities, it does not seem possible that the prescribed age limit be lower than that required for 'dangerous and unhealthy' occupations." Furthermore, the European Convention on the Exercise of Children's Rights, adopted by the Council of Europe on 26 Janury 1996, sets the age of majority at 18 years.

International humanitarian law provides for the protection of defined groups, such as civilians. Those who may legitimately kill and be killed are the members of armed forces. The recruitment of under-18s makes this group of children lawful objects of attack because they are members of armed forces.

Goodwin-Gill and Cohn support the choice of 18 as the minimum age for recruitment and participation in hostilities on the basis that "participation in the political process is ... a reasonably accurate indicator of the moment at which the community as political body recognises the intellectual maturity of the individual",[400] and show that the vast majority of countries in all regions of the world have set 18 years as the voting age. They also cite state practice as supporting 18 years as the minimum age for recruitment. Of the 185 states they surveyed, only seven are quoted as having a lower age for compulsory recruitment, and only six a lower voting age. They argue that "Given the essentially political dimension to armed conflict, whether national or non-international, the choice of 18 as the moment of

---

[399] Comments on the Report of the Working Group on a draft optional protocol to the Convention on the Rights of the Child on involvement of children in armed conflicts, contained in UN document E/CN.4/1996/WG.13/2 of 23 November 1995, p 12. However, child soldiering may be included in the proposed ILO Convention, due to be adopted in 1999, prohibiting the involvement of under-18s in the worst forms of child labour.

[400] G. Goodwin-Gill and I. Cohn: Child Soldiers (Clarendon Press, Oxford, 1994), p 7

transition to adulthood does not seem unreasonable. Indeed, in principle, it would seem wrong to condemn the unenfranchised to die as a consequence of political decisions on which they can exercise no influence." Furthermore, they argue that the international human rights and humanitarian law prohibition of the death penalty on those under 18 years at the time of the commission of the offence, applicable alike in times of peace and war, "acknowledges the reduced ability of those under 18 to appreciate the nature of their action in the context of criminal responsibility. The same consideration, however, is not given to the capacity of the child or young person to evaluate the reasons for death in combat."[401]

The physical, emotional and social impact of involvement in armed conflict of those under 18 years is too well documented in other chapters of this book to warrant repetition here. Equally it is clear that the distinction between voluntary and compulsory recruitment is often not as clear cut in reality as it might seem in principle, and that the effect of participation in any case remains the same. Although voluntary recruitment into peacetime armed forces may seem innocuous, the fact remains that if the armed forces become engaged in a conflict, these recruits are likely to be used and are, in any case, legitimate objects of attack. Furthermore, the evidence of this book is that children are recruited predominantly because not enough adult recruits are forthcoming, or in order to use them as spies or to commit atrocities. These are not reasons compatible with the Convention on the Rights of the Child. In addition, the active participation of some children exposes other children, particularly in the conflict zones, to pressure to join one or other side, and to suspicion of involvement making them liable to attack, interrogation and other harassment. In addition to reducing the actual participation of under-18s, reducing the suspicion of such participation should help to protect all children in affected areas from attack, harassment and detention. In a conflict, once child recruitment starts, it tends to escalate rapidly and the age of recruitment is driven down. Therefore, the only effective way to protect children from being involved in hostilities is to prevent their recruitment.

[401] Ibid, p 9

CHAPTER 8

Strategies to Prevent Recruitment

EXAMPLES HAVE ALREADY BEEN given of tactics used to avoid forcible recruitment and by families to distance their children and thus dissuade them from volunteering or spontaneously taking part. Essentially, however, such tactics do nothing to tackle the phenomenon of child recruitment, addressing only the question of who is affected with, as discussed in Chapter 3, certain systematic consequences. There are, however, instances where action has been taken which has had a more general effect in discouraging child recruitment. Furthermore, this book's analysis of how and why children are recruited and which groups of children are most vulnerable to recruitment enables some consideration of general preventive strategies. However, although there are common issues, specific strategies in particular situations must be informed by an understanding of why children in that situation are actively involved, and which children are most vulnerable to recruitment ("risk mapping").

The general strategies are considered in this section by reference to the main actors whose assistance may be enlisted. However, the sections are not self-contained and, in particular, the role of different actors, including those of international organisations and specialised agencies, are woven into the sections rather than being treated under a separate heading.

ARMED FORCES AND ARMED OPPOSITION GROUPS

Since most of the recruitment is carried out directly or indirectly by the armed forces and armed opposition groups themselves, without their cooperation it will not be possible to eliminate child recruitment. Obviously those who deliberately choose to make use of child soldiers are not in the short term going to be amenable to persuasion, but the evidence does not suggest that such cases are in the majority. More often, the recruitment of children seems to be the consequence of negligence and inadequate procedures rather than of conscious desire. Where governments, and armed opposition groups, have proper recruitment procedures, children are less likely to become involved. Furthermore, it is quite clear from all the evidence that if the recruitment of children is accepted, or not opposed, it will continue. It is, therefore, imperative that where it is known to take place, it is exposed and pressure (public or private) is brought to bear on the perpetrators and that local communities and others in a position to oppose and to take preventive steps are encouraged and supported.

As shown above, there is a body of informed military opinion which holds that young recruits are at best a mixed blessing, at worst a positive liability. In purely military terms there is no clear evidence of any particular effectiveness of child soldiers in general; quite a lot suggests the contrary. If this message could be spread more widely, all sorts of armed forces might be encouraged to consider improvements to their recruiting methods, not just to avoid international opprobrium, but with the more direct result of improving their own efficiency by eliminating the enlistment of unsuitable material.

In addition, enlightened self-interest points towards the non-use of children by all parties in view of the emerging evidence that the widescale involvement of child combatants is in itself a factor in the continuation of armed conflicts, in addition to their propensity to commit atrocities and the links to violent criminal activity. The mutuality of interests of the opposing forces in this regard reinforce the idea that local agreements on a minimum age of recruitment are a valid option, and one which communities, other countries, the Special Representative of the UN Secretary-General (SRSG) for Children in Armed Conflict, international and national organisations and non-governmental organisations should encourage. Further, such agreements, if implemented, would have a beneficial effect for all children, particularly those in or from the conflict zones, by reducing the direct pressures on them and by reducing the general level of suspicion of them and, therefore, of harassment and other ill-treatment. Communities have a direct and particular interest in encouraging such agreements.

GOVERNMENTS

The self-evident action which governments can take is to outlaw the recruitment of children in all government armed forces and other armed groups which they establish, condone or permit to bear arms, for example local militias/civil defence forces, and to introduce proper recruitment procedures. In practice, this cannot be effective unless accompanied by a ban on all forms of forced recruitment. Many of the practices described in this report amount to violations of fundamental human rights (arbitrary detention, disappearance, inhuman and degrading treatment and summary executions) which governments are in any case under a duty to prevent and remedy. It is, however, facile to imagine that the promulgation of legislation will affect the situation on the ground unless accompanied by legal remedies and institutions strong enough and willing to tackle abuses.

> ...on the subject of this illegal procedure, 596 reports for May and June 1995 have been received about youths who have been recruited by force. As a result of intervention and pressure from this institution, 333 people have managed to be set free, 148 of these being under 18. All this in spite of the ineffective intervention of some judges...[402]

Apart from the question of recruitment methods themselves, governments can play a major part in preventing recruitment into both their own armed forces and armed opposition groups. They can seek to insulate the civilian population from the physical and economic consequences of the conflict. Keeping schools open, or establishing, renovating and reopening schools,[403] not only gives the young a civilian focus but avoids the development of the

---

[402] Report of the Human Rights Ombudsman quoted in the case study for Guatemala (translated from the Spanish)
[403] Case studies for Lebanon and Liberia

situation whereby the gap is filled by institutions linked to armed opposition groups. The availability of schooling for refugee and displaced children, including adolescents, also needs to be addressed. Other governments, international and national organisations and non-governmental organisations can help in terms of advocacy and of provision.

Recognition of conscientious objection to military service and provision for alternative civilian service help to counter the overwhelming militarisation of society by validating alternatives. The establishment of the necessary procedures to enable this, will also assist in the general regularisation of recruitment procedures and the requirement to ensure that potential recruits know their rights and how to claim them.

Again, quite apart from preventing it from taking place, under-age recruitment is hard to detect and monitor where individuals lack documentary proof of age. Article 7(1) of the Convention on the Rights of the Child, which has achieved near universal ratification, requires states parties to provide proper birth records. Governments could do much to deter under-age recruitment by thorough and universal implementation of this obligation, including for refugees and internally displaced children. Donor governments, international and national organisations and non-governmental organisations could play a major part in encouraging and assisting in the establishment of systematic procedures for universal, routine registration of births. However, even if such procedures were implemented immediately, registration of existing children also needs to be addressed, and the replacement of documentation lost in transit for refugee and displaced children. Where the question of birth registration is controversial because it is linked to the question of nationality (for example, in relation to some refugee and displaced populations or where the parents are stateless), the two issues should be treated separately to ensure that births continue to be documented.

Separate from the problem of establishing age, age itself may not be considered the relevant factor in determining adulthood or eligibility for military service, but rather the criterion may be height, physical development, weight or other factors. Since Governments establish laws and accept international legal obligations relating to a specific age, they also have a duty to engage with their own populations, including military and civilian leaders and the children themselves, to develop an understanding of why age is significant and of the effects on children of involvement in armed conflict, in addition to seeking means to enforce the application of relevant ages. For children themselves, knowing their rights and that age is a protective factor are crucial since many may have no idea that there are rules governing the minimum age of recruitment.

Governments can also play a part in putting pressure on armed opposition groups in relation to their conduct by becoming parties to the Statute of the International Criminal Court and to relevant international humanitarian law treaties and applying them in internal armed conflicts, thus making such groups also accountable under humanitarian law. Other governments can also bring pressure to bear on such groups (including not supplying arms to them) and those which are parties to the Convention on the Rights of the Child and relevant international humanitarian law treaties have an obligation to "ensure respect" for international humanitarian law.

Governments who are hosting refugee populations can assist by ensuring that armed elements are not present in nor have access to camps and that camps are not situated close to international borders, thus encouraging the actual involvement of refugees in cross-border fighting, or the perception of their involvement which is likely to result in attacks on the camps, with consequent likelihood of the militarisation of the children. Furthermore, ensuring productive educational activities for refugee and displaced children and adolescents (with the assistance of UNHCR, UNICEF, international and national NGOs and others) is a major factor in reducing volunteering into armed groups. More generally, the application of the Guiding Principles on Internal Displacement would assist in addressing many of the problems of internally displaced persons which feed the causes of child militarisation.

At an even more fundamental level, as long as armed conflicts occur (particularly internal ones), children are likely to become involved in them; the longer they continue, the greater and the younger the child participation tends to become. These may not be the best arguments for governments to address the causes of conflicts and to bring them to an early end. However, the point has to be made that all the other strategies combined will not by themselves solve the problem of child soldiers.

One obvious step would seem to be that it is in the self-interest of governments to regulate the production and flow of lightweight automatic weapons and other weapons which are light and simple enough for children to use to such devastating effect, whether in armed conflicts or in violent crime. As a matter of urgency, Governments should encourage these issues to be addressed by international, regional and sub-regional bodies, since to be at all effective controls will need to be multifaceted.

COMMUNITIES, NATIONAL AND INTERNATIONAL NON-GOVERNMENTAL ORGANISATIONS

Sometimes the attitude of the local community may be enough to dissuade illegal forced recruitment. One case study reports a claim:

> *that the Army has only changed its recruitment methods in some places where the inhabitants have opposed it ... However, in areas where they have not met any opposition they continue to use forced recruitment.*[404]

Another reports "the notable decline in the number of 'levies' [army recruiting drives] in the areas under the jurisdiction of the parish churches which denounced this activity."[405]

A few individual cases of forced recruitment of minors have been remedied through *habeas corpus* proceedings brought by the national Committee for the Defence of Human Rights, or by families with its support but such proceedings cannot effect structural change.[406]

The populations affected – ethnic groups, or mothers of actual or potential recruits – have in some places formed organisations which have tried to bring political pressure to bear against governmental recruitment of children by writing letters, organising demonstrations, appealing to parliament and the army leadership, public denunciation of forced recruitment, receiving testimonies of mothers whose sons have been taken, and even going on hunger strike.[407] If the children can prove that they are under

[404] Case study for Guatemala (translated from the Spanish)
[405] Case study for Peru (translated from the Spanish)
[406] Case study for Honduras
[407] Case study for El Salvador, Guatemala and Paraguay

A boy playing with a toy gun in South Africa 1994. Counteracting the glorification of war and violence is one way of preventing children's participation in armed conflicts. Photo: Gamma/IBL.

age according to the national law, they are quite often returned to their communities, especially if their parents or guardians claim them and the human rights non-governmental organisations protest applying a great deal of pressure on the army.[408]

Outside organisations, too, can wield influence. One instance is recorded where protests from aid organisations led to the return of boys and men forcibly recruited from a refugee camp,[409] and there are reports elsewhere of the negotiation of agreements on non-recruitment of under-15s by humanitarian organisations with local commanders.[410] Such organisations can also help by informing children of their right not to be recruited and helping to counter some of the pressures to join the armed struggle.[411]1Much more, however, could be done by international non-governmental organisations (in particular, children's and humanitarian ones in the field) and by UNICEF, UNHCR and the Special Representative of the Secretary-General for Children in Armed Conflicts (SRSG) through monitoring and reporting on abuses, raising the issues with those in authority, supporting local groups working on the issues, and educating children and young people, and the military and other authorities. In addition, all field programmes should include "risk mapping" – identifying in that particular situation the children who are being recruited, by whom and why. For example, are they in a particular area of concentration of the fighting, of a particular age, gender, type, ethnic group. In this way, specific planning can be made to try to protect these groups over and above general programming, for example, to ensure an adequate standard of living for child-headed households.

By contrast to all this, the greatest succour which can be given to systematic forced and child recruitment is the silence of popular opinion and international organisations, however good the excuses for such silence:

---

[408] Case study for Guatemala

[409] Case study for Burma/Myanmar

[410] SPLM/Operation Lifeline Sudan Agreement on Ground Rules, July 1995

[411] Human Rights Watch/Africa, Children's Rights Project: Children of Sudan (September 1995), p 85

The most active Human Rights Organisations ... both admitted having helped mothers to demand their forcefully recruited sons, but neither had done anything to discourage or prevent minors of age from recruiting voluntarily. Their representatives admitted that in this respect lots of violations of Rights of Children were committed, but said apologetically that they had so much work to do on those days, so many more urgent challenges ... most [citizens] seem to have felt that fighting the war was not so very different from the other duties and jobs children traditionally perform among the poor sectors of society. Children endure all kinds of hardships, and are deprived in so many ways that their participation in the armed forces did not attract any major attention by the society and its institutions. Anyhow, the great majority of children on both sides of the conflict were children that never really "counted/mattered" in the eyes of society ... a concept prevails that in the war situation the constitution and international conventions lose their validity.[412]

FAMILIES

The maintenance of families and the attitude of those families are one of the key factors in reducing child recruitment. Those parents who see the military struggle as "barbaric and uncivilised", or are opposed because of their religious background, or "because of their experiences about the devastating effect of war",[413] will do everything they can to ensure the non-involvement of their children. In some cases this may include assigning sons to the monkhood for varying lengths of time in order to protect them from recruitment.[414] Similarly protected are those who

come from families where the family relationships are stable and where parents have positive and healthy attitudes toward life itself; where, for the parents, the worth of their children's lives is above their own national, patriotic or political interests.[415]

[412] Case study for El Salvador
[413] Case study for Liberia
[414] Case studies for Burma/Myanmar and Cambodia
[415] Case study for the former Yugoslavia

It might seem that the families already do what they can, particularly in the matter of adopting evasive strategies. Many children, however, fail to benefit from such protection simply because they have no family background. The importance of supporting and maintaining family structures cannot, therefore, be overemphasised: not only physically, but also economically, socially and emotionally. Family reunification programmes for street children, orphaned and abandoned children[416] can reduce the pool of ready recruits. Making maximum efforts to prevent children becoming separated from their families during displacement or on return, and ensuring early family tracing and reunification. For those children without families, for example orphans, trying to provide foster or substitute care to provide the kind of protection normally associated with the family, including access to adequate food, housing and security. Governments, agencies and non-governmental organisations can all play a part in such endeavours.

Where children, and their families, know the national and international law concerning the age of recruitment and participation in hostilities, children are less likely to be recruited in the first place. It may also enable them to challenge illegal recruitment, particularly where there are governmental institutions able and willing to bring legal cases.[417] Banding together to form mutual support groups or campaigning organisations increases the likelihood of success and of effecting changes in policies and practices. Ignorance of the national and international laws means that this factor is not even considered by the children themselves or by their families.[418] Informing children of their rights, including non-recruitment of anyone under 15 years by any party, may enable unaccompanied children to withstand pressure to join. Too often even if they know the law, families do not believe in its efficacy in a conflict situation (with justification), and if children (or these particular children) are not valued by society, their involvement will be harder to prevent.[419] Outside support, from international organisations, the Special Representative of the Secretary General for

[416] Case study for Liberia
[417] Case study for Guatemala and Honduras
[418] Case study for the Philippines
[419] Case study for El Salvador

Children in Armed Conflicts (SRSG) or NGOs, for groups who are trying to enforce the law may make a significant difference in their success rate.

It is clear that the attitude of families and communities is a crucial factor both in the immediate situations and the longer term. As long as the involvement of children is accepted as inevitable, the situation is unlikely to change fundamentally. On the other hand, where communities oppose such involvement and they have a modicum of power or influence, the incidence of such involvement is negligible. The fact that the overwhelming picture is of the poorest, least educated and most disadvantaged children being the recruits undermines any argument of a lower age for recruitment or participation being culturally acceptable, as does the existence of national legislation which does not take account of such supposed cultural variations. The preventive strategies employed by families reinforce the distinction between functional families with a measure of education, income and influence who can act to protect their own children from compulsory recruitment and volunteerism, and dysfunctional, illiterate and poor families, or children without families, who are unable to exercise such choices and strategies. Not only are they, therefore, more at risk initially, but the absence of the first group leaves them under greater pressure. The need for systematic protection for all children cannot, therefore, be overemphasised.

CHAPTER 9

Recommendations

LEGAL ISSUES

- International law should recognise 18 years as the minimum age for recruitment (compulsory or voluntary) into any kind of armed forces and armed groups and for any kind of participation in hostilities. The proposed Optional Protocol to the Convention on the Rights of the Child on involvement of children in armed conflict should establish these standards and be adopted as soon as possible.

- Support should be given to the international coalition established to inform public opinion about the situation of child soldiers and the need for the Optional Protocol to raise the minimum recruitment and participation ages to 18 years.

- National coalitions should be established to lobby governments who currently recruit below this age to raise their voluntary and compulsory recruitment ages to at least 18.

- Where a Government derogates under a human rights treaty, it should accept the applicability of at least common Article 3 of the Geneva Conventions of 1949.

- Governments should become parties to Additional Protocol II to the Geneva Conventions of 1949, and governments and opposition groups should be encouraged to apply the Protocol.

- African states should become parties to the 1990 African Charter on the Rights and Welfare of the Child, which sets 18 years as the minimum age for recruitment and participation in hostilities.

- The follow-up report to the analytical study of the UN Secretary-General on Fundamental Standards of Humanity (E/CN.4/1998/87) should continue to explore the issues of the standards applicable in internal armed conflicts and internal violence and of the accountability of non-state actors under international law.

- Governments should become parties to the statute of the International Criminal Court which includes in the category of war crimes in both international and non-international armed conflicts, any form of recruitment of under-15s and their use in hostilities.

- Support the inclusion of a prohibition on child soldiering in the proposed ILO Convention on the worst forms of Child Labour to be adopted in 1999.

THE REGULATION OF RECRUITMENT:

- All governments and armed opposition groups who currently have persons under 18 years in their armed forces should be urged to demobilise them immediately, and to refrain from recruiting under-18s in future. In current conflict/internal violence situations, all parties should be urged to make local agreements to this effect. The mutual nature of such agreements should encourage parties to keep to them, and in addition to reducing the actual participation of under-18s, by reducing the suspicion of such participation should help to protect all children in the affected area from harassment, detention or pre-emptive recruitment. All governments, as well as relevant international and regional organisations, the Special Representative of the Secretary General for Children in Armed Conflict (SRSG), specialised agencies and international and national non-governmental organisations can assist in advocating for, negotiating and monitoring such agreements.

- Governments should introduce proper recruitment procedures for their armed forces:

(1) Where conscription/compulsory recruitment exists, it must be given a legal basis, with a clear minimum age, which should be not less than 18 years. The law should be publicised both to those

recruiting and to the public at large in ways which will ensure that those liable to conscription and those not liable to it, know their rights and duties. Proper safeguards, including requirement of proof of age, and military disciplinary or criminal penalties for infringements, should be available and enforced to ensure that under-age recruitment does not take place.

(ii) The role of local militias/civil defence forces as recruiting agents should be eliminated. Their recruitment ages and procedures, and those of any armed group established, condoned, armed or permitted to bear arms by the government, should be brought into line with those for regular government armed forces, and the provisions of UN Commission on Human Rights resolution 1994/67 implemented. This resolution recommends: "(a) Civil defence forces shall only be deployed for the purpose of self-defence; (b) Recruitment into them shall be voluntary and shall be effectively controlled by public authorities; (c) Public authorities shall supervise their training, arming, discipline and operations; (d) Commanders shall have clear responsibility for their activities; (e) Civil defence forces and their commanders shall be clearly accountable for their activities; (f) Offences involving human rights violations by such forces shall be subject to the jurisdiction of the civilian courts". Effective, practical measures to ensure the implementation of these recommendations require further consideration.

(iii) Where there is conscription/compulsory recruitment, provision should be made for exemption from military service on grounds of conscience and for an alternative, non-military, non-punitive, service for conscientious objectors to military service which is compatible with the reasons for the objection, in accordance with UN Commission on Human Rights resolution 1998/77.

(iv) Voluntary recruitment should require appropriate safeguards both in relation to age and to voluntariness with the obligation on the recruiter to obtain confirmation of the full and informed consent of the individual and of any persons (such as parents) with legal responsibility for the individual volunteering.

OTHER PREVENTIVE MEASURES:

- The routine provision of official birth records is an essential pre-requisite for the prevention of under age recruitment into government armed forces (whether compulsory or voluntary). It is also required by Article 7(1) of the Convention on the Rights of the Child.

(i) Governments should, as a matter of priority, establish effective and universal systems for registering births and providing the necessary documentation.

(ii) UNICEF, other agencies (including UNHCR in relation to persons of concern to them) and non-governmental organisations should assist governments in establishing such systems.

(iii) In the meantime, and for those born before such systems are fully operational, UNICEF, UNHCR, other agencies and non-governmental organisations should seek to assist individuals/families in establishing ages and ensuring the provision of appropriate documentation.

(iv) Issues about nationality should not be allowed to stand in the way of registration of births.

- UNICEF, UNHCR, other agencies and non-governmental organisations should inform themselves about the age(s) for compulsory and/or voluntary recruitment under the relevant national law, and ensure that:

(i) it is in conformity with international law;

(ii) persons with whom they are in contact know their rights – including the minimum age(s) relevant, and exempted categories if any – and what recourse is available to them if they are unlawfully recruited;

(iii) the law is applied in practice, and if not, should take this up with the appropriate national authorities, the Committee on the Rights of the Child and other UN mechanisms such as the Special Representative of the Secretary General for Children in Armed Conflict (SRSG), UN Commission on Human Rights Special

Rapporteurs and Working Groups on Summary Executions, Arbitrary Detention, Torture, Disappearances, Sale of Children, and relevant Country Rapporteurs.

- The most effective ways for Governments to reduce volunteerism into opposition armed groups are:

(i) Not to attack or subject to violent harassment, children themselves, their families and homes.

(ii) Where children are involved in armed opposition groups, to treat them in accordance with recognised international standards on capture, and also with the Convention on the Rights of the Child, and allow regular access to all child detainees by the International Committee of the Red Cross.

(iii) To provide access to education and vocational training for all children.

(iv) To address the economic, social and political causes of the conflict.

- UNHCR and other agencies and non-governmental organisations should make provision, in particular in the form of education and vocational training, for children and adolescents in order to reduce volunteerism, including for refugees and internally displaced.

- The Committee on the Rights of the Child, the Special Representative of the Secretary General for Children in Armed Conflict (SRSG) and the thematic and country specific mechanisms of the UN Commission on Human Rights, should specifically monitor, respond to and report on violations of the rights of children who are soldiers.

- The civilian nature and humanitarian character of refugee and displaced persons camps should be ensured. Armed elements should not be allowed in or have access to such camps and they should be sited away from international borders.

- The Special Representative of the Secretary General for Children in Armed Conflict (SRSG) should continue to undertake field visits, meet with all parties to armed conflicts and secure commitments

not to recruit or use children, and to institute monitoring and follow-up mechanisms in relation to such commitments and provisions
of international human rights and humanitarian law.

• UN agencies, national and international NGOs should make use of
the ARC training module on child soldiers to train their staff on
how to prevent the recruitment of refugee and displaced children,
and about demobilisation and reintegration of child soldiers.[420]

ISSUES REQUIRING FURTHER CONSIDERATION:

• The World Health Organisation should undertake a study of the
short- and long-term physical and mental health aspects of child soldiering, taking account of the specific situation of adolescents and
also of gender issues.

• The UN Sub-Commission on Prevention of Discrimination and
Protection of Minorities should undertake a study of the treatment
of recruits in government armed forces with a view to examining its
compatibility with human rights standards, as well as the consequences of inhuman and degrading treatment for the recruits and
for the subsequent behaviour of the armed forces.

• The Committee on the Rights of the Child and the UN Commission
on Crime Prevention and Criminal Justice should consider the
questions of applicability to children of:

(i) emergency and anti-terrorism legislation;

(ii) military law, punishment and discipline (including in military
schools), and

(iii) should also consider the question of the age of criminal responsibility, in particular in relation to situations of armed conflict,
states of emergency and terrorism, and their compatibility with the
Convention on the Rights of the Child. Governments should review
their legislation to ensure its conformity with the Convention and
other international standards.

[420] See Annex 5 for more information.

188

- Governments, regional, sub-regional and international bodies should consider how to best regulate the production and flow of light-weight automatic weapons, and other small arms.

- The UN Commission on Human Rights Special Rapporteur on Mercenaries should study the possible linkage between recruitment as children and subsequent involvement as mercenaries.

- The UN Commission on Crime Prevention and Criminal Justice should consider the question of possible linkage between recruit-ment as children and subsequent involvement in violent crime.

DEMOBILISATION, REHABILITATION AND SOCIAL REINTEGRATION

- At the time of demobilisation, and to ensure the rehabilitation and reintegration of child combatants into civil society, the provisions of the Convention on the Rights of the Child should constitute the guiding principles in all actions to protect their rights and wel-fare.

(i) In particular, the international community must advocate govern-ments and other concerned groups to ensure that the participation of children in conflict is recognised in peace agreements and re-lated documents, in order that their needs are not neglected and are incorporated as a matter of principle in plans for demobilisation, rehabilitation and social reintegration.

(ii) Advocacy and action on behalf of child soldiers should be im-plemented with due regard for the children's right to privacy, should respect their dignity, protect their physical and emotional well-being, and ensure that no children are exploited with the purpose of drawing attention to any particular situation/programme. Agencies involved with the welfare of child soldiers should establish guidelines and procedures that reflect professional ethical stan-dards to address these issues.

(iii) The children should be informed of solutions that are being con-sidered to address their circumstances and, with due regard for their age and stage of maturity, their participation should be incorporated into defining strategies for their rehabilitation and social reintegration.

- When it is known that children are to be demobilised, international organisations, in particular UNICEF and UNHCR, and international and local non-governmental organisations should implement programmes of family tracing and reunification as a matter of urgency, to ensure the return of children to their communities of origin.

(i) Alternative systems of care that are appropriate to the culture and circumstances of the children should be established to meet the needs of children who, for whatever reason, cannot be reunited with their families. The institutionalisation of children should be discouraged, and only implemented when all alternative forms of care have been considered.

(ii) All such programmes should be implemented in partnership with the appropriate governmental authorities to ensure that there is a capacity to meet the needs of the children, and that their well-being is monitored.

- The international community, in particular UNICEF and the non-governmental organisations in partnership with local organisations, should ensure that communities are not left alone to bear the burden of rehabilitation and social reintegration of the children, and should recognise the crucial role they can play in this process.

(i) Programmes that incorporate the participation of the children, their families and communities, and enable them to normalise their daily lives and prevent further recruitment of under-age combatants should be implemented. In this regard, projects directed at implementing sustainable initiatives that provide support to the family and community and improve the economic and social conditions that influenced the initial recruitment of the children should be considered a priority.

(11) Programmes of education and vocational training should also be a priority concern in the rehabilitation and social reintegration of child soldiers. Alternatives to traditional educational systems will need to be considered, and additional training for teachers made available. Opportunities for education and vocational training must be appropriate to the circumstances of the communities, and not lead to unrealistic aspirations on the part of the children. For older children, or those who due to long periods of time spent with the military have difficulty in returning to school, the creation of opportunities for self-employment to compensate for their lack of skills should be considered.

(111) The needs of child soldiers should be addressed within overall planning and service provision for all children affected by armed conflict, to ensure that they are not marginalised within their communities. However, special programmes to assist the social reintegration of especially vulnerable groups (e.g. girl soldiers; disabled child combatants) may be required in some circumstances. Appropriate methods of coping with these situations should be determined through dialogue with the children, their families and communities.

- Governments should ensure that follow-up of demobilised child soldiers takes note of community attitudes towards the culpability of children who were actively involved in conflict, and ensure that they are awarded due protection within the national juvenile justice system, and in accordance with the provisions of the Convention on the Rights of the Child. International and local non-governmental organisations should assist in these procedures, and intervene where necessary to protect the rights and well-being of the children, and to promote reconciliation and understanding of the causes and consequences of children's participation in conflict.

CHAPTER 10

Executive Summary

THE APPLICABILITY OF THE UN CONVENTION ON THE RIGHTS OF THE CHILD

This book focuses on the causes and consequences of the involvement of children in armed forces and armed opposition groups and as combatants in armed conflict. The principles of action to effect change for children who have been recruited or have participated in armed conflict and to promote their best interests must be guided by their needs and those of their families and communities as described in the Convention on the Rights of the Child. The provisions of the Convention, to which nearly every state is now a party, should constitute the guiding principles in all actions to protect the rights and welfare of child recruits and combatants. Where children are involved in government armed forces or armed opposition groups they should be treated in accordance with recognised international standards on capture, and also in accordance with the provisions of the Convention. The Convention should be the basis to ensure that the participation of children in conflict is recognised in peace agreements and related documents, and that their needs are incorporated in plans for demobilisation, rehabilitation and social reintegration. (Chapters 4, 5)

THE CONSEQUENCES OF PARTICIPATION IN CONFLICT FOR CHILDREN

The principal consequence for the vast majority of children who participate in armed conflict is separation from their families and communities.

This separation occurs in violent circumstances at a period in their lives when children have most need of the care and support that family life provides. Both government forces and armed opposition groups should not attack or subject to violent harassment children themselves, their families and homes. (Chapters 2, 3, 5)

As members of both government forces and armed opposition groups, children's deaths are the result of active involvement in combat where their inexperience and, frequently, lack of training results in high casualty rates. They are commonly used on the front lines, and their size and agility lead them to be given particularly hazardous assignments. Suicide missions are considered the particular province of adolescents. Many children who are injured in combat are left to die from their wounds, or are shot. Those who are too weak to keep up with the group, who attempt to escape, either to avoid recruitment or to desert, are usually executed. Children die from injuries incurred during beatings, whether as a punishment or to break the spirit of new recruits. They are also more prone to die from starvation and preventable diseases contracted in the unhygienic conditions in which they live. (Chapters 2, 4)

With the exception of certain armed forces and armed opposition groups whose treatment of children is more humane, the treatment and training of child soldiers involves a high degree of risk for their physical well-being, especially for the youngest amongst them. Their bodies are still developing and they are thus at greater risk of injury and disability from the privations that are common in military life. These include poor diet, insanitary conditions and inadequate health care, and the rigours of harsh training routines and excessive punishments that can leave them weakened and debilitated. (Chapter 4)

There are numerous instances of children being routinely administered drugs and alcohol, particularly before a battle. Sexual abuse of both boys and girls is not infrequently reported, entailing for many the risk of sexually transmitted diseases, HIV/AIDS, and pregnancy for the girls. Where the pregnancy is terminated in insanitary conditions, this can put the girls at additional risk of associated disorders. (Chapters 3, 4)

The most frequent injuries suffered by child soldiers are loss of limbs, loss of hearing and blindness. These disabilities will impose additional hardships in the future, compromising their chances to take advantage of educational and vocational programmes, and impeding their social reintegration as they become an additional burden for an already impoverished family. (Chapters 4, 5)

Over and above the specifically physical consequences the children suffer, they are also subjected to degrading and humiliating treatment to subordinate them to authority. Such treatment can influence their psychological well-being, destroying their self-esteem and leading to violent solutions to problems. Some children commit suicide following such events. (Chapter 4)

Given these consequences of recruitment and participation in conflict and the fact that the recruitment of children into armed opposition groups often fuels under-age recruitment into government armed forces and *vice versa*, the recruitment of children should stop. In addition, the active participation of some children exposes other children, particularly in the conflict zones, to pressure to join one or other side, and to suspicion of involvement making them vulnerable to attack, interrogation and other harassment. All governments and armed opposition groups who currently have persons under 18 years in their armed forces should be urged to demobilise them immediately, and to refrain from recruitment of under-18s in future. In current conflict/internal disturbance situations, all parties should be urged to make local agreements to this effect. The mutual nature of such agreements should encourage parties to keep to them and, in addition to reducing the actual participation of under-18s, by reducing the suspicion of such participation, should help to protect all children in the affected area from attack, harassment and detention. All governments, as well as relevant international and regional organisations, specialised agencies and international and national non-governmental organisations can assist in advocating for and negotiating such agreements. (Chapters 4, 8)

DEMOBILISATION, REHABILITATION AND
SOCIAL REINTEGRATION

The demobilisation of child soldiers is seen as essentially the preliminary phase in a process of rehabilitation and social reintegration. Such a process cannot be accomplished without recognition of the participation of children in conflict. This is a requirement of local communities, government and armed opposition groups involved in the conflict, and the international community. It is only when children are demobilised that the reality of their experiences in conflict and the consequences for their well-being can be recognised and addressed. (Chapter 5)

The effects of participation in conflict cannot be explained in straightforward terms. They are principally described as child welfare concerns that have an impact on the children's immediate social and cultural environment, influence both physical and emotional well-being, and limit their opportunities for education, vocational/skills training and, ultimately, their chances for employment. In almost all instances their experiences as child soldiers are described as having a negative effect on these varied aspects of their lives that would normally contribute to their healthy development and as compromising their capacity to reintegrate into their families and communities. (Chapter 5)

The family is considered the children's primary resource, and therefore family tracing and reunification, and the association of children to their communities of origin, should be implemented as a matter of urgency when it is known that children are to be demobilised. (Chapter 5)

Re-attachment to family and community will determine to a large extent effective social reintegration. As a result of the conflict, however, the majority of families are impoverished. Thus programmes addressing rehabilitation and social reintegration should encourage the participation of families and communities. They should also incorporate support to their families and communities, not only to the children. These should be directed at implementing sustainable initiatives that improve the economic and social conditions that in most cases influenced the initial recruitment of the children. (Chapter 5)

Resources available to meet children's needs, both during and after conflict, are limited and there is little likelihood of resources being directed specifically at child soldiers. In any event, all children in the affected area have suffered deprivations due to the war. Thus a more effective and equitable approach will be to address the needs of former child soldiers within overall planning and service provision for all children affected by

196

armed conflict. This would reduce any risk of marginalising children who have been directly involved in combat. A similar consideration will apply for those children who, for whatever reason, cannot be reunited with their families. For them, alternative systems of care must then be considered that meet the children's needs, which are culturally acceptable and, again, avoid any possibility that they will be marginalised. The institutionalisation of children should be discouraged, and only implemented when all alternative forms of care have been considered. (Chapter 5)

However, some children will be especially vulnerable, due to particular experiences and events. Most evident will be children whose disabilities pose a financial burden for families already experiencing economic hardship. Also, children who have been seriously affected by their experience of traumatic events may require special assistance. Another group for whom social reintegration may prove difficult is girl soldiers with children. Whilst special programmes to assist the social reintegration of vulnerable groups should be considered, dialogue with their families and communities should determine appropriate methods of coping with these situations, in particular addressing the consequences of "traumatic events" and other conflict-related problems. (Chapter 5)

The case studies made repeated reference to the linkage between education, employment opportunities, and the economic security of the children's families in determining successful social reintegration, and preventing re-recruitment. Without education, children's future prospects for employment are limited, and the army may be their only opportunity to earn money and contribute to the family economy. Education is also the system within which children's lives can be normalised. Through meaningful and productive activity they can be helped to overcome their experiences and develop an identity separate from that of the soldier. Programmes of education and vocational training should, therefore, be seen as a priority concern in the rehabilitation and social reintegration of child soldiers. Given the potential of large numbers of demobilised child soldiers in some situations, their varying ages, experiences and their own attitudes towards education, alternatives to traditional educational systems will need to be considered, and additional training for teachers made available. Opportunities for education and vocational training must be appropriate to the circumstances of the communities, and not lead to unrealistic aspirations on the part of the children. For older children or those who, due to long periods of time spent with the military, have difficulty in returning to school, the creation of opportunities for self-employment to compensate for their lack of skills should be considered. (Chapter 5)

The children should be informed of solutions that are being considered to address their circumstances and, with due regard for their age and stage of maturity, their participation should be incorporated into defining strategies for their rehabilitation and social reintegration – see Article 12 of the Convention on the Rights of the Child. (Chapter 5)

Communities cannot be left alone to bear the burden of rehabilitation and social reintegration. Of principal concern is that there are implemented programmes of assistance which have long-term sustainability for these most vulnerable victims of armed conflict, and for their families and communities who have suffered extreme deprivation as a result of the participation of their children in the conflict. Organisations directly involved with the welfare of children affected by participation in armed conflict – principally UNICEF, but also UNHCR and international and local non-governmental organisations – should identify procedures to address the worst effects of participation in armed conflict, and move towards involving the children, their families and communities in processes to effect peaceful strategies. (Chapter 5)

PREVENTING RECRUITMENT OF CHILDREN

The main reason why children are recruited into government armed forces is not because children as such are wanted but because of shortage of manpower and inadequate recruitment procedures. Recruitment of children particularly occurs when forced recruitment, such as press-ganging, is practised, when large numbers of soldiers are required and when the conflict is prolonged. Governments can prevent or minimise under-age recruitment by introducing proper recruitment procedures and prohibiting forced recruitment. Where conscription/compulsory recruitment occurs, it must be given a legal basis, with a clear minimum age, which should be not less than 18 years. The law should be publicised both to those recruiting and to the public at large in ways which will ensure that those liable to conscription, and those not liable to it, know their rights and duties. Proper safeguards, including requirement of proof of age, and military disciplinary or criminal penalties for infringements, should be available and enforced to ensure that under-age recruitment does not take place. (Chapter 2)

To require proof of age presupposes that individuals have a birth registration or identity document which can be used to demonstrate their age. Without such a document, genuine problems about under-age recruitment arise, but the non-availability of proof also encourages unscrupulous disregard of age requirements. The routine provision of official birth records is required by Article 7 of the Convention on the Rights of the Child. It is an essential prerequisite for the prevention of under-age recruitment (whether compulsory or voluntary) into government armed forces. Governments should, as a matter of priority, establish effective and universal systems for registering births and providing the necessary documentation. UNICEF, other agencies (including UNHCR in relation to persons of concern to them), national and international non-governmental organisations and donor governments, can assist in establishing such systems. In the meantime, and for those born before such systems are fully operational, governments, UNICEF, UNHCR, other agencies and non-governmental organisations should seek to assist individuals/families in establishing ages and ensuring the provision of appropriate documentation. (Chapters 2, 8)

Where there is provision for both conscription and voluntary recruitment with a lower entry age for volunteers, under-age conscripts can be passed off as volunteers or pressured into volunteering. Not only should clear minimum ages be established, but the age for volunteers should be the same as for conscripts. In any case, procedures are required to ensure that volunteers are just that. Governments should establish appropriate safeguards for voluntary recruitment both in relation to age and to voluntariness with the obligation on the recruiter to obtain confirmation of the free, full and informed consent of the individual and of any persons (such as parents) with legal responsibility for the individual volunteering. (Chapters 2, 8)

In some countries recruitment for government armed forces is carried out indirectly by local militias/civil defence forces or others. Such indirect systems of recruitment are prone to abuse, particularly when combined with the requirement to produce a specified number of recruits, rather than named individuals. The role of local militias/civil defence forces as recruiting agents for government armed forces should be eliminated. The recruitment ages and procedures for all armed militias/civil defence forces themselves should be aligned with those for regular government armed forces and with UN Commission on Human Rights resolution 1994/67, including the requirement that all recruitment into them be voluntary. (Chapter 2)

There is a need for children themselves and for their families to be informed about their rights and how to enforce them, and for government recruitment practices to be monitored by national and international bodies, to ensure conformity with international law in both law and practice. A lack of national and international response has permitted widespread child recruitment. Governments should include specific information on their recruitment laws and practices in their reports to the Committee on the Rights of the Child. UNICEF, UNHCR, other agencies and national and international non-governmental organisations should inform themselves about the age(s) for compulsory and/or voluntary recruitment under the relevant national law, and ensure: (a) that it is in conformity with international law; (b) that persons with whom they are in contact know their rights – including the minimum age(s) relevant, and exempted categories if any – and what recourse is available to them if they are unlawfully recruited; (c) that the law is applied in practice and, if not, should take this up with the appropriate national authorities, the Committee on the Rights of the Child and other UN mechanisms such as the Special Representative of the Secretary General for Children in Armed Conflicts (SRSG), the UN Commission on Human Rights Special Rapporteurs and Working Groups on Summary Executions, Arbitrary Detention, Torture, Disappearances, Sale of Children, and relevant Country Rapporteurs. (Chapter 8)

The Special Representative of the Secretary General for Children in Armed Conflicts (SRSG), the Committee on the Rights of the Child, thematic and country-specific mechanisms of the UN Commission on Human Rights should systematically monitor, respond to and report on violations of the rights of children who are soldiers including their recruitment, treatment in armed forces and on capture. (Chapter 8)

Families can help to counter societal and peer group pressure on children to join armed forces or armed opposition groups. Society itself can also help by reducing the militarisation of daily life and providing alternative models to the glorification of war and military leaders. Specifically, provision should be made for exemption from military service on grounds of conscience and an alternative, non-military and non-punitive form of service for conscientious objectors to military service, in accordance with UN Commission on Human Rights resolution 1998/77. (Chapters 2, 8)

Children become soldiers primarily because they are available, have less power than adults to oppose recruitment and are, therefore, more likely to be forced, intimidated or persuaded into joining government armed forces or armed opposition groups. Child recruitment falls unequally. Those most likely to be recruited, whether willingly or unwillingly, are: (a) first and foremost, children separated in any way from their families; (b) the economically and socially deprived (the poor, whether rural or urban, those without access to education, vocational training, and a reasonable standard of living); (c) other marginalised and disenfranchised groups (such as street children, certain minorities, refugees and the internally displaced); and (d) children from the conflict zones themselves. The factors leading to child recruitment usually occur in combination. Governments, UNICEF, UNHCR, other agencies and organisations, national and international non-governmental organisations, separately and together, all have roles to play in maintaining or reuniting families, providing access to education and vocational training for all children, ensuring adequate economic provision so that such access is real rather than theoretical and that children have the prospect of a livelihood (Convention on the Rights of the Child Articles 27 and 28). The particular needs and vulnerability of separated children, refugees and internally displaced children and families, need to be addressed. (Chapters 2, 3)

Children and their families need to know their rights, including the minimum ages for legal recruitment into government armed forces and that no-one under the age of 15 years should be in any armed force or armed group or participating in hostilities, and to whom they can complain or appeal (Convention on the Rights of the Child, Article 38). Governments, UNICEF, UNHCR (in relation to persons of concern to them), national and international non-governmental organisations should assist in the protection of all children by ensuring that they know their rights in this regard and the means of redress open to them. (Chapter 8)

The single major factor for children volunteering into armed opposition groups is their personal experience of attack, ill-treatment or harassment of themselves or their families by government armed forces. Sometimes they join armed opposition groups in search of revenge but more often it is a desire for protection because of their sense of vulnerability. Governments can reduce the incidence of children volunteering into armed opposition groups by: (a) not attacking or subjecting to violent harassment, children themselves, their families and homes; (b) where children are involved in armed opposition groups, treating them in accordance with recognised international standards on capture, in particular the Convention on the Rights of the Child, and allowing regular access to all child detainees by the International Committee of the Red Cross. (Chapters 2, 4)

An international standard of 15 years as the minimum age for recruitment into any armed force or armed group and for participation in hostilities is too young (Convention on the Rights of the Child, Article 38, 1977 Additional Protocol I, Article 77(2), 1977 Additional Protocol II, Article 4(3)(c)). Most national law already sets the minimum age for conscription at 18 years (as well as the age of political majority)[421]. International law also recognises 18 as the age of adulthood in many respects. Article 1 of the Convention on the Rights of the Child sets the general age of majority at 18 years. The special status of persons under 18 years is recognised both in relation to the prohibition on the application of the death penalty[422] and in relation to employment or work "likely to jeopardise the health, safety or morals of young persons"[423]. In relation to recruitment, Article 38(3) marks out the 15–18 year age group and exhorts states recruiting in that age bracket to give priority to the eldest. The evidence of this report is that children are recruited predominantly because not enough adult recruits

[421] See Annex 1
[422] Convention on the Rights of the Child, Article 37(a), International Covenant on Civil and Political Rights, Article 6(5), Additional Protocol I, Article 77(5), Additional Protocol II, Article 6(4)
[423] International Labour Organisation Convention No. 138 on Minimum Age, 1973, Article 3(1)

are forthcoming, or in order to use them as spies or to commit atrocities. These are not reasons compatible with the Convention on the Rights of the Child. In addition, the evidence of the physical, mental and emotional effects on under-18s of involvement in armed forces or armed opposition groups and participation in hostilities clearly point to the need for the minimum age to be raised. The proposed Optional Protocol to the Convention on the Rights of the Child on involvement of children in armed conflicts should be adopted as soon as possible setting 18 years as the minimum age for any kind of recruitment into any armed forces or armed groups, and for any kind of participation in hostilities. (Chapters 5, 7)

LEGAL AND OTHER ISSUES

Child soldiers are usually treated in the same way as all other soldiers. However, in government armed forces, the treatment of recruits is often inhuman and degrading in the extreme. Such treatment may particularly affect children because they are less robust physically, mentally and emotionally. The general issue of the ill-treatment of recruits and its conformity with international human rights standards needs to be addressed. The UN Sub-Commission on Prevention of Discrimination and Protection of Minorities should undertake a study of the treatment of recruits in government armed forces, with a view to examining its compatibility with human rights standards, as well as the consequences of inhuman and degrading treatment for the recruits and for the subsequent behaviour of the armed forces. (Chapter 4)

Under civilian legal regimes, special provision is normally made at all stages in the legal process to take account of the age of the child. In situations of armed conflict or internal disturbances, often emergency or anti-terrorism legislation is introduced which takes no account of age. Alternatively the age of criminal responsibility is reduced in the face of children's involvement. Similarly, military law, punishment and discipline do not take the impact on children into account. The Committee on the Rights of the Child should consider the questions of applicability to children of (a) emergency and anti-terrorism legislation and (b) military law, punishment and discipline (including in military schools), and should also consider the question of the ages of criminal responsibility in particular in relation to situations of armed conflict, states of emergency and terrorism, and their compatibility with the Convention on the Rights of the Child. Governments should review their legislation to ensure its conformity with the Convention and other international standards. (Introduction, Chapter 4)

In many situations of internal armed conflict, internal strife or disturbances, the government denies that there is an armed conflict, thereby denying the applicability of international humanitarian law to itself and any opposing armed groups, and derogating from its obligations under international human rights law. (a) Where a Government derogates under a human rights treaty, it should not be able to deny the applicability of at least common Article 3 of the Geneva Conventions of 1949. (b) Governments should become parties to Additional Protocol II to the Geneva Conventions of 1949, and governments and opposition groups should be encouraged to apply the Protocol. (Introduction, Chapter 7)

The ready availability of simple-to-operate lightweight automatic weapons has transformed the capacity of children to serve as combatants on something approaching an equal footing with adults. The factors which combine to produce child soldiers — disruption of families, lack of educational and economic opportunities – are exacerbated by the children's participation in armed forces or armed opposition groups. On leaving such service, the lack of other opportunities open to them, together with the acquired skills in weapons, encourages some to seek a future in violent crime. Governments should regulate the production and flow of lightweight automatic weapons. (Chapter 1)

ANNEX I

Child Soldiers
in the World

SOURCES:

Unless another reference is given, the information in this annex comes from three sources, as follows:

Rädda Barnen database, as at April 1998: for all information on child soldiers and strengths of armed forces and groups (original sources and updated information may be obtained by consulting the database at *http://www.rb.se)*, also whether there is an ongoing conflict, voting ages, and ratifications to the Geneva Conventions and Protocols. It should however be noted that the figures for the total numbers of child soldiers have generally been estimated for the purposes of this table and are not attributable to Rädda Barnen.

Guinness World Fact Book, (Guinness Publishing, London, 1994): for populations and duration of compulsory military service.

Goodwin-Gill, G. & Cohn, I., *Child Soldiers,* (Oxford University Press, 1994): for the existence of conscription and for legal recruitment ages.

Among other sources consulted were:

War Resisters International (Horeman, B., Stolwijk, M. & Luccioni, A.), *Refusing to bear arms: a world survey of conscription and conscientious objection to military service;* Part I: Europe, (London, 1997)

Jongman, A. J. & Schmid, P. *Contemporary Conflicts* and *World Conflict Map 1994–95,* in PIOOM Newsletter and Progress Report, Vol. 7, No. 1 (Leiden University, 1966), pp 14–23.

Project Ploughshares, *Armed Conflicts Report 1995* and *Armed Conflicts Report 1996* (Institute of Peace & Conflict Studies, Waterloo, Ontario).

Sollenberg, M. & Wallenstein, P. *Major Armed Conflicts*, in SIPRI Yearbook 1995 (Oxford University Press, New York), pp 26–36.

United Nations Document E/CN.4/1997/99 *The Question of Conscientious Objection to Military Service; Report of the Secretary-General prepared pursuant to Commission Resolution 1995/83*, presented to the UN Commission on Human Rights, January, 1997.

GENERAL NOTES AND EXPLANATIONS

(NB: an asterisk in the table indicates that the entry is expanded or explained in the footnotes on page 226 et seq.)

Blank spaces in this table mean that no information is to hand. They should not be read as zero.

Population (Column B)

Figures are generally estimates for 1993 and 1992. To minimise inconsistency, these have not been updated in a piecemeal fashion even when later figures have become available.

Armed Forces, Paramilitaries, Armed Opposition Groups (Column C)

All acronyms used indicate armed opposition groups; if the names of Government-sponsored forces are quoted they are spelled out in full. Acronyms are explained in Annex 4 on page 263.

Published figures of the strength of different countries' armed forces vary considerably between countries and between sources depending on which reserve and paramilitary-type forces are included in the figure. In this table reserves are excluded and wherever possible figures for the strength of paramilitary forces and militias are separated from those for the regular armed forces; within these categories, however, different forces are not listed unless there is further information of particular relevance.

Armed opposition groups are as a rule mentioned only if we can quote an estimate of their total strength, and/or there have been specific allegations that they include minors.

It will be appreciated that whereas some of the figures for the manpower of government armed forces are exact, other figures in this column, particularly relating to opposition armed groups, are estimates of varying degrees of crudity.

Child Soldiers – Total Numbers (Column D)

The entries in this column are the compiler's best guesses on the basis of the information available and do not always report or agree with any total cited in the source. Where there is a wide range of estimates of both the size of force and the percentage of child soldiers, a middle estimate for the number of child soldiers is here suggested; where the information is altogether too sketchy, an indication is given simply of whether the total is likely to be in thousands (?000), hundreds (?00) etc. Where there have been substantial recent demobilisations of child soldiers (ie. since 1994/5, when the material for the Case Studies was gathered), the figures given relate to the pre-demobilisation situation.

Girls (Column F)

Where a figure is quoted, this is the percentage of child soldiers who are female.

Lowest age recorded (Column G)

If the existence of child soldiers is indicated but no figure is given in this column, the minimum recruitment age is also the actual minimum age reported. It will be understood that even assuming totally accurate sources, the ages quoted are not necessarily typical or still valid.

Conscription (Column I)

Where two figures are given for the length of compulsory military service, the shorter is usually for service in the army, the longer for certain other services (often the navy). Periods of full-time service only are included, excluding any refresher courses or time in the reserves.

"Selective conscription" exists where not all male citizens are obliged either to perform compulsory military service or to establish grounds for exemption. In some countries female citizens too are subject to conscription.

Legal Recruitment Ages (Columns J and K)

The age quoted under "Conscription" is the age at which citizens become liable to compulsory military service; that under "Volunteers" the minimum age, where known, at which recruits may legally be accepted. In many cases, this represents the age at which those who will be liable for compulsory military service may opt for personal reasons to satisfy the requirement before they are obliged to. In some cases, conversely, a conscription age is given where the enabling legislation is in place although conscription does not at present apply. It is quite normal for compulsory registration for military sevice to be required at a year or more below the age of liability; this sometimes leads to a misleadingly low age being quoted. On the other hand, in many countries military service is not in practice performed at the formal minimum age of liability.

Treaties (Column M)

This column indicates whether each state is party to the 1949 Geneva Conventions (GC), Additional Protocol I (API) and Additional Protocol II (APII). The Convention on the Rights of the Child (CRC), which also contains provisions on the recruitment of children and their participation in hostilities, has now been ratified by all the states listed except Somalia, Taiwan and the United States of America.

The following parties to the Convention on the Rights of the Child have no armed forces and have therefore been omitted from the table: Andorra, Dominica, Grenada, the Holy See, Iceland, Kiribati, Liechtenstein, the Maldives, the Marshall Islands, Micronesia (Federated States of), Monaco, Nauru, Palau, Saint Kitts & Nevis, Saint Lucia, Saint Vincent & the Grenadines, San Marino, the Solomon Islands, Tuvalu, and Western Samoa.

A	B	C	D	E	F
Country	Population (millions)	Armed Forces Paramilitaries and/or Militias Armed Opposition Groups	Child Soldiers Total Number	As % of force	Girls?(%)
AFGHANISTAN	18.1				
Taliban		25,000	10,000	?	No
various factions		240,000	108,000	* 45	no report
ALBANIA	3.4	54,000			
paramilitary		13,500			
opposition groups (see note)		not known	?00	?	no report
ALGERIA	25.9	124,000			
paramilitary forces		45,000			
Communal Guards		100,000	?000	?	
AIS etc.		> 2,000	?00	?	no report
GIA		not known	?00	?	no report
ANDORRA	0,06	none	-	-	-
ANGOLA	10.5	110,500	* 3,500	* 3	
Reaction Police		15,000			
UNITA		10,000–60,000	* 7,200	* 12	
Cabindan opposition groups		6,000	?00	?	Yes(30–40)
ANTIGUA & BARBUDA	0.1	150			
ARGENTINA	33.5	73,000			
ARMENIA	3.6	60,000			
paramilitary		1,000			
AUSTRALIA	17.6	57,400	* 925	2	Yes (18) *
AUSTRIA	7.8	* 38,000	* <100	<1	
AZERBAIJAN	7.3	66,700			
paramilitary forces		40,000			
Karabakh opposition groups		20,000–25,000	?000	?	
BAHAMAS	0.3	860			
BAHREIN	0.5	11,000			
BANGLADESH	115.1	121,000	?00	?	
paramilitary forces		49,500			
JSS / SB		5,000			
BARBADOS	0.3	610			
BELARUS	10.4	81,800			
BELGIUM	10.0	44,450			
BELIZE	0.2	1,050			
BENIN	5.1	4,800			
BHUTAN	*1.4	* 11,000	>500	* 5	No*
paramilitary forces		* 10,000			
militias		* >2,200	>200	* 10	
BOLIVIA	7.7	33,500			
BOSNIA& HERZEGOVINA	*4.4	40,000	see note		
BOTSWANA	1.3	7,500			
BRAZIL	156.5	314,700			
BRUNEI DARUSSALAM	0.3	5,000			
BULGARIA	8.5	101,500			
militias		* 100,000			

Lowest Age Recorded	Armed Conflict?	Conscription and Duration of Service (in months)	Legal Recruitment Ages		Voting Age	Treaties CRC GC API APII	
			Conscripts	Volunteers			
13 *10	Yes	see note	see note		?	CRC/GC	AFGHANISTAN
*10	Yes *	Yes (12) *	* 19		18	CRC/GC/API-II	ALBANIA
not stated 15 16	Yes	Yes (6)	19		18	CRC/GC/API-II	ALGERIA
-	-	-	-	-	18	CRC/GC	ANDORRA
8 8 8	Yes	Yes	18		18	CRC/GC/API	ANGOLA
	No	No *			18	CRC/GC/API-II	ANTIGUA & BARBUDA
	No	No *	* 18		18	CRC/GC/API-II	ARGENTINA
	No *	Yes (24)	?		18	CRC/GC/API-II	ARMENIA
	No	No *	* 18	* 17	18	CRC/GC/API-II	AUSTRALIA
	No	Yes (8) *	18	17	19	CRC/GC/API-II	AUSTRIA
14	Yes *	Yes (24) see note	? * 17		21	CRC/GC	AZERBAIJAN
	No	No			18	CRC/GC/API-II	BAHAMAS
	No	No		18	18	CRC/GC/API-II	BAHREIN
	Yes *	No		* 16	18	CRC/GC/API-II	BANGLADESH
	No	No			18	CRC/GC/API-II	BARBADOS
	No	Yes (18)	?		18	CRC/GC/API-II	BELARUS
	No	No		18	18	CRC/GC/API-II	BELGIUM
	No	No			18	CRC/GC/API-II	BELIZE
	No	Selective *	?		18	CRC/GC/API-II	BENIN
* 11 not stated	No	Selective (1/3)	18	16	18	CRC/GC/API-II	BHUTAN
	No	Yes (12) *	* 19		18	CRC/GC/API-II	BOLIVIA
	Yes *	Yes *	* 18		18	CRC/GC/API-II	BOSNIA& HERZEGOVINA
	No	No			22	CRC/GC/API-II	BOTSWANA
	No	Yes (12)	19		18	CRC/GC/API-II	BRAZIL
	No	No			?	CRC/GC/API-II	BRUNEI DARUSSALAM
	No	Yes (18) *	* 18		18	CRC/GC/API-II	BULGARIA

A Country	B Population (millions)	C Armed Forces Paramilitaries and/or Militias Armed Opposition Groups	D Child Soldiers Total Number	E As % of force	F Girls?(%)
BURKINA FASO	9.8	* 10,000			
BURMA/MYANMAR	44.6	429,000	>50,000	* 10–66	Yes
People's Police Force		50,000			
People's Militia		35,000	?000	?	
ABSDF		2,000	> 200	* >10	
KA		> 1,000	250	* 25	
KNLA		4,000	1,000	* 25	Yes *
MTA / UWSA		20,000	5,000	* 25	
BURUNDI	5.6	< 30,000	* 2,000	?	no report
gendarmerie		3,500			
opposition groups		not known	?00	?	Yes
CAMBODIA	9.3	* 140,500	6,000	* 4	
Khmer Rouge opposition		* <9,000	2,000	* 25	
CAMEROON	13.1	13,000			
gendarmerie		9,000			
CANADA	28.1	61,600	* 250	< 1	
paramilitary forces		11,100			
CAPE VERDE	0.3	2,650			
CENTRAL AFRICAN REPUBLIC	3.0	4950			
gendarmerie		2,300			
CHAD	6.1	30350			
gendarmerie		4,500			
opposition groups		not known			
CHILE	13.5	94,300	?00	?	
Carabinieros		31,200			
FPMR / MIR		1,300			
CHINA (now including Hong Kong)	1164.1	2840,000			
People's Armed Police		* 800,000			
COLOMBIA	34.0	146,300	* 4756	4	
National Police Force		87,000	* 5,000	6	
FARC/ELN/EPL		9,000–23,000	2,500	* 10 – 30	Yes
paramilitaries *		"several thousand"	2,000	15–50	
COMOROS	0.5	* 900			
MPA		not known			
CONGO BRAZZAVILLE		10,000			
paramilitary forces		5,000			
COBRA		not known	?00		
CONGO, DEMOCRATIC REPUBLIC	41.1	* 20,000–40,000	?000	?	
ADFL			*10,000-15,000		
COSTA RICA (paramilitary only)	3.2	7,000			
COTE D'IVOIRE	12.5	* 7,100			
paramilitary		* 7,800			
CROATIA	4.8	58,000	?00	?	
paramilitary police		40,000			
CUBA	10.7	* 125,000–135,000	50,000	?	
Interior Ministry Forces		65,000			

Lowest Age Recorded	Armed Conflict? (H)	Conscription and Duration of Service (in months) (I)	Legal Recruitment Ages (J-K)		Voting Age (L)	Treaties CRC GC API APII (M)	
			Conscripts	Volunteers			
	No	No			18	CRC/GC/API-II	BURKINA FASO
* 7	Yes	No		18	18	CRC/GC	BURMA/MYANMAR
* 12							
* 11							
* 10							
* 11 / 12							
	Yes	No		16	18	CRC/GC/API-II	BURUNDI
* 12							
* 8	Yes	Yes (36/42) *	* 18	18	18	CRC/GC/API-II	CAMBODIA
* 5							
	Yes *	No			20	CRC/GC/API-II	CAMEROON
	No	No		16	18	CRC/GC/API-II	CANADA
	No	Yes (24) *	?		18	CRC/GC/API-II	CAPE VERDE
	Yes	Selective *	?		18	CRC/GC/API-II	CENTRAL AFRICAN REPUBLIC
	Yes	Selective *	?		18	CRC/GC	CHAD
	No	Yes (24)	18	16	18	CRC/GC/API-II	CHILE
	Yes *	Selective (36/48)	18		18	CRC/GC/API-II	CHINA
	Yes	Selective (12/24)	* 18	* 16	18	CRC/GC/API-II	COLOMBIA
* 8							
* 8							
	Yes				18	CRC/GC/API-II	COMOROS
	Yes	No			18	CRC/GC/API-II	CONGO BRAZZAVILLE
14							
* 12	Yes	Yes *	* 18		* 18	CRC/GC	CONGO, DEMOCRATIC REPUBLIC
8							
	No	No	* 18		18	CRC/GC/API-II	COSTA RICA
	No	Selective *	21		21	CRC/GC/API-II	COTE D'IVOIRE
	Yes *	Yes (10) *	* 16	* 17	18	CRC/GC/API-II	CROATIA
	No	Yes (36) *	* 16		16	CRC/GC/API-II	CUBA

A	B	C	D	E	F
Country	Population (millions)	Armed Forces Paramilitaries and/or Militias Armed Opposition Groups	Child Soldiers		
			Total Number	As % of force	Girls?(%)
CYPRUS	0.8	10,000			
Turkish Cypriot forces		* 4,000			
CZECH REPUBLIC	10.3	* 62,750			
DENMARK	5.2	* 34,000	* 38	* <1	Yes (11)*
Home Guard		* 64,600	* <25	* <1	
DJIBOUTI	* 0.7	* 18,000			
DOMINICAN REPUBLIC	7.6	24,500			
ECUADOR	11.0	57,100			
EGYPT	57.1	450,000			
Ministry of Interior forces		232,000			
opposition groups		not known	?00	?	no report
EL SALVADOR	5.5	28,400	?000	?	
paramilitary police		12,000			
EQUATORIAL GUINEA	0.4	1,320			
paramilitary		* 2,000			
MAIB		> 300			
ERITREA	3.7	46,000			
opposition groups		>4,000			
ESTONIA	1.5	3,500	<100	?	
Border Guard		2,000			
ETHIOPIA	51.8	* 120,000	?00	?	No ? *
opposition groups		* >22,000	* >8,000	* >35	
FIJI	0.8	3,600			
FINLAND	5.0	31,000	<100	?	
Frontier Guard		3,400			
FRANCE	57.5	380,820			
gendarmerie		* 95,700			
GABON	1.3	4,700			
paramilitary forces		* 4,800			
GAMBIA	1.0	200			
armed police force		* 600			
GEORGIA	5.5	33,200			
opposition groups		7,000			
GERMANY	79.8	347,100	?00	?	
Bundesgrenzenschutz (border guard)		24,500	?00	?	
coast guard		550			
GHANA	15.6	7,000			
paramilitary forces		6,000			
GREECE	10.3	162,300	?00	?	
paramilitary forces		30,500			
GUATEMALA	9.7	40,700	* > 1,000	?	no report
paramilitary forces		12,300			
PAC militias *		* 800,000	* ?000	?	no report
URNG *		* 800–1,100	* ?00	?	Yes *

Lowest Age Recorded	Armed Conflict?	Conscription and Duration of Service (in months)	Legal Recruitment Ages		Voting Age	Treaties CRC GC API APII	
			Conscripts	Volunteers			
	No	Yes (26) Yes (24) *	18 * 18		21	CRC/GC/API-II	CYPRUS
	No	Yes (12) *	* 19		18	CRC/GC/API-II	CZECH REPUBLIC
	No	Selective (4/13) *	* 18	* 18 * 18	18	CRC/GC/API-II	DENMARK
	No	No			18	CRC/GC/API-II	DJIBOUTI
	No	No *			18	CRC/GC/API-II	DOMINICAN REPUBLIC
	No	Selective (12)	19	18	18	CRC/GC/API-II	ECUADOR
see note	Yes *	Selective (36)	18		18	CRC/GC/API-II	EGYPT
	No	Yes *	?	16	18	CRC/GC/API-II	EL SALVADOR
	No	Yes	18		?	CRC/GC/API-II	EQUATORIAL GUINEA
	No	Yes (18) *	* 18		?	CRC	ERITREA
	No	Yes (9/12) *	* 18	* 17	18	CRC/GC/API-II	ESTONIA
12 not stated	Yes *	Selective (12/18)	18		21 18	CRC/GC/API-II	ETHIOPIA
	No	No			21	CRC/GC	FIJI
	No	Yes (8/11) *	* 18	* 17	18	CRC/GC/API-II	FINLAND
	No	Yes (10)*	* 18		18/23*	CRC/GC/API-II	FRANCE
	No	No			21	CRC/GC/API-II	GABON
	No	No		* 18	18	CRC/GC/API-II	GAMBIA
	Yes	Yes (24)	?		18	CRC/GC/API-II	GEORGIA
	No	Yes (10) *	* 18	* 17 * 16	18	CRC/GC/API-II	GERMANY
	Yes	No			18	CRC/GC/API-II	GHANA
	No	Yes (19/23) *	* 18	16	18	CRC/GC/API-II	GREECE
* 12 * 11 * 12	Yes *	No *		15	18	CRC/GC/API-II	GUATEMALA

215

A	B	C	D	E	F
Country	Population (millions)	Armed Forces Paramilitaries and/or Militias Armed Opposition Groups	Child Soldiers		
			Total Number	As % of force	Girls?(%)
GUINEA	7.4	9,700			
paramilitary		* 9,600			
GUINEA BISSAU	1.0	9,250			
GUYANA	0.7	1,600			
paramilitary		* 3,500			
HAITI	6.9	3,000			
HONDURAS	5.2	18,800	* >1,000	?	
Public Security Forces		5,500			
opposition guerrilla groups *		not known	* <100	?	Yes *
HUNGARY	10.3	* 55,000			
INDIA	897.6	1145,000	?000	?	
paramilitary		1088,000			
Kashmiri opposition groups		3,000	?00	?	
Manipur and Tripura opp. groups		26,000–31,000	?000	?	Yes
INDONESIA	188.2	283,000	?000	?	
paramilitary		160,000			
militias		1,500,000			
opposition groups		> 500			
IRAN	60.8	518,000			
Basij		* 200,000	?000	?	No
Law-Enforcement Forces		150,000			
NLA		15,000–30,000			
KOP *		8,000			
IRAQ	19.5	387,500	?000	?	
paramilitary forces		45,000–55,000			
KDP *		* 40,000			
PUK		* 32,000	?000	?	
other opposition groups		4,500			
IRELAND	3.5	12,700			
ISRAEL/OCCUPIED TERRITORIES	6.5	175,000			
Border Police		6,050			
Palestinian security forces		35,000			
PNLA *		8,000			
HAMAS		300	<100	?	
other Palestinian groups		>5,000			
ITALY	57.1	325,150			
paramilitary forces		255,700			
JAMAICA	2.5	3,320			
JAPAN	124.7	235,600			
JORDAN	3.3	104,050	* 1,850	<2	
KAZAKHSTAN	17.2	35,100			
KENYA	28.1	24,200			
paramilitary police		5,000			
KOREA, NORTH	22.6	1055,000			
KOREA, SOUTH	44.0	672,000			

Lowest Age Recorded	Armed Conflict?	Conscription and Duration of Service (in months)	Legal Recruitment Ages		Voting Age	Treaties CRC GC API APII	
			Conscripts	Volunteers			
	No	Yes (24)	?	19	18	CRC/GC/API-II	GUINEA
	No *	Yes *	* 18		18	CRC/GC/API-II	GUINEA BISSAU
	No	No *			18	CRC/GC/API-II	GUYANA
	Yes *	No *		18	18	CRC/GC/API-II	HAITI
* 13 * 15	No	No *		17	18	CRC/GC/API-II	HONDURAS
	No	Yes (9) *	* 18	* 18	18	CRC/GC/API-II	HUNGARY
11 11	Yes *	No *	* 18	* 16	18	CRC/GC/API-II	INDIA
	Yes	Selective (24)	?	17	17	CRC/GC	INDONESIA
* 9	Yes	Yes (24) *	* 18	no limit?	15	CRC/GC	IRAN
10	Yes	Yes (24) *	19	* 15	18	CRC/GC	IRAQ
	No	No *		* 17	18	CRC/GC	IRELAND
12	Yes	Selective(24/36)*	* 18	17	18	CRC/GC	ISRAEL/OCCUPIED TERRITORIES
	No	Yes (10) *	* 18	* 17 * 18	18/25	CRC/GC/API-II	ITALY
	No	No			18	CRC/GC/API-II	JAMAICA
	No	No *		* 18	20	CRC/GC	JAPAN
	No	No	* 18	* 17	19	CRC/GC/API-II	JORDAN
	No				18	CRC/GC/API-II	KAZAKHSTAN
	Yes	No			18	CRC/GC/API-II	KENYA
	No	Yes (36/120) *	* 18		17	CRC/GC/API	KOREA, NORTH
	No	Yes (30/36) *	* 20	* 18	20	CRC/GC/API-II	KOREA, SOUTH

A	B	C	D	E	F
Country	Population (millions)	Armed Forces Paramilitaries and/or Militias Armed Opposition Groups	Child Soldiers		
			Total Number	As % of force	Girls?(%)
KUWAIT	1.4	15,300			
KYRGYSTAN	4.4	7,000			
LAOS	4.5	29,000	?000	?	
militia		>100,000			
UNLNF		< 2,000			
LATVIA	2.6	* 17,000			
LEBANON *	2.8	55,100	?00	?	
Internal Security Force		13,000			
SLA		2,000–2,500	?00	?	Yes *
Hezbollah		3,000–5,000	?00	?	Yes *
Palestinian fringe groups		500	<100	?	Yes *
LESOTHO	1.9	2,000			
LIBERIA	2.9	* 9,000	* none?		
various factions		* 59,000	* 12,800	* 25	Yes (1) *
LIBYA	4.6	65,000	?00	?	
LITHUANIA	3.8	5,250			
National Guard *		* 12,500			
LUXEMBOURG	0.4	800	?	?	
gendarmerie		560			
MACEDONIA	2.1	15,400			
MADAGASCAR	13.4	21,000			
MALAWI	10.6	5,000			
MALAYSIA	19.1	111,500			
MALI	8.6	7,350			
paramilitary *		* 7,800			
MALTA	0.4	1,950			
MAURITANIA	2.2	15,650	?00	?	
paramilitary forces		5,000			
MAURITIUS (paramilitary only)	1.1	1,800			
MEXICO	90.0	175,000	?00	?	
Rural Defence Militia		14,000			
EZLN		>3,000	?00	?	
MOLDOVA	4.4	11,030			
MONGOLIA	2.3	11,000			
MOROCCO (& WESTERN SAHARA)	26.5	196,300	1,200	?	
paramilitary forces *		* 40,000			
Polisario (opposition)		3,000–6,000	?00	?	
MOZAMBIQUE	15.7	11,000–34,000			
NAMIBIA	1.4	5,800	?00	?	
NEPAL	19.4	46,000			
paramilitary police		40,000			
NETHERLANDS	15.3	* 57,200	* 140	< 1	Yes (17)*
NEW ZEALAND	3.4	9,550	* 300	* < 3	Yes (11)*
NICARAGUA	4.3	17,000	?00	?	
Frente Norte (opposition)		1,200			

G	H	I	J	K	L	M	
Lowest Age Recorded	Armed Conflict?	Conscription and Duration of Service (in months)	Legal Recruitment Ages Conscripts	Volunteers	Voting Age	Treaties CRC GC API APII	
	No	Yes (24)	18	18	21	CRC/GC/API-II	KUWAIT
	No	? *			18	CRC/GC/API-II	KYRGYSTAN
	Yes *	Yes (18)	15		18	CRC/GC/API-II	LAOS
	No	Yes (18) *	* 19	* 18	18	CRC/GC/API-II	LATVIA
* 17	Yes	Yes	18	18	21	CRC/GC	LEBANON
* 9							
15							
* 10							
	No	No			21	CRC/GC/API-II	LESOTHO
	Yes	see note			18	CRC/GC/API-II	LIBERIA
* 6							
	Yes	Yes (36/48)	* 18	* 17	18	CRC/GC/API-II	LIBYA
	No	Yes (12) *	* 19	* 18	18	CRC/GC	LITHUANIA
	No	No		17	18	CRC/GC/API-II	LUXEMBOURG
	No	Yes (9) *	* 18		18	CRC/GC/API-II	MACEDONIA
	No	Yes (18) *	?		18	CRC/GC/API-II	MADAGASCAR
	No	No		18	18	CRC/GC/API-II	MALAWI
	No	No			21	CRC/GC	MALAYSIA
	No	Selective (24) *	?		18	CRC/GC/API-II	MALI
	No	No			18	CRC/GC/API-II	MALTA
	No	Yes (24) *	?	16	18	CRC/GC/API-II	MAURITANIA
	No	No		* 18	18	CRC/GC/API-II	MAURITIUS
	Yes *	Yes (12) *	* 18	* 16	18	CRC/GC/API-II	MEXICO
12							
	No	Yes (12) *	?		18	CRC/GC/API-II	MOLDOVA
	No	Yes (24)	18		18	CRC/GC/API-II	MONGOLIA
not stated	No	Yes (18)	18		20	CRC/GC	MOROCCO
17		see note					
	Yes *	Selective (24)	18		18	CRC/GC/API	MOZAMBIQUE
	No	Yes	16		18	CRC/GC/API-II	NAMIBIA
	Yes *	No		18	18	CRC/GC	NEPAL
	No	No *	* 17	* 17	18	CRC/GC/API-II	NETHERLANDS
	No	No		* 17	18	CRC/GC/API-II	NEW ZEALAND
	No	No		17	16	CRC/GC	NICARAGUA

A	B	C	D	E	F
Country	Population (millions)	Armed Forces Paramilitaries and/or Militias Armed Opposition Groups	**Child Soldiers**		
			Total Number	As % of force	Girls?(%)
NIGER	8.5	5,300			
paramilitary forces		5,400			
NIGERIA	88.5	77,000			
Port Security Police		2,000			
NORWAY	4.3	33,600	* < 100	* <1	
Home Guard		* 80,000	* 1,700	* 2	Yes (20)*
OMAN	2.2	43,500			
PAKISTAN	128.0	587,000	?000	?	
paramilitary		247,000			
PANAMA (paramilitary only)	2.6	11,800			
PAPUA NEW GUINEA	3.9	4,300	?00	?	
BRA		1,000	<100	?	
PARAGUAY	4.6	20,200	* 9,000	* 45	No *
paramilitary police		14,800	* 2,400	* 16	No *
PERU	22.6	125,000	?000	?	Yes
paramilitary		78,000			
Rondas Campesinas militias		100,000	?000	?	Yes
Sendero Luminoso (opposition) *		* 1,500	?00	?	Yes
MRTA		200	?	?	Yes
PHILIPPINES	65.0	110,500			
paramilitary forces		102,500			
NPA		8,000	* 1,000–2,000	* 10–30	Yes *
MILF, MNLF, etc.		12,000–40,000	?000	?	
POLAND	38.3	241,750	?000	?	
paramilitary forces		23,400			
PORTUGAL	10.4	59,300	<100	?	
National Republican Guard		* 20,900			
Public Security Police		* 20,000			
Border Guard		* 8,900			
QATAR	0.5	11,800			
ROMANIA	22.8	226,950			
RUSSIAN FEDERATION	148.0	1,240,000	?00	* <1	
paramilitary forces		583,000			
Chechen opposition		not known	?000	?	see note
RWANDA (see note)	7.5				
RPF *		* 20,000			
former Government forces		* 30,000	? 100	* <1	
militias		not known	?000	?	no report
SAO TOME & PRINCIPE	0.1	* 900			
SAUDI ARABIA	15.3	105,500			
paramilitary forces		* 85,500			
SENEGAL	7.9	13350			
gendarmerie		4,000			
SEYCHELLES	0.1	400			
paramilitary		* 800			

Lowest Age Recorded (G)	Armed Conflict? (H)	Conscription and Duration of Service (in months) (I)	Legal Recruitment Ages (J/K)		Voting Age (L)	Treaties CRC GC API APII (M)	
			Conscripts	Volunteers			
	Yes	Selective (24)	?		18	CRC/GC/API-II	NIGER
	Yes *	No		18	?	CRC/GC/API-II	NIGERIA
	No	Yes (6/12) *	* 18	* 17 * 16	18	CRC/GC/API-II	NORWAY
	No	No			?	CRC/GC/API-II	OMAN
	Yes	No		* 16	21	CRC/GC/API-II	PAKISTAN
	No	Yes *	?		18	CRC/GC/API-II	PANAMA
not stated not stated	Yes *	No			18	CRC/GC/API-II	PAPUA NEW GUINEA
* 12 * 12	No	Yes (18/24)	18	18	18	CRC/GC/API-II	PARAGUAY
* 11 * 9 * 9 * 11	Yes	Selective (24)	* 18	* 16	18	CRC/GC/API-II	PERU
  * 10 12	Yes	Yes	18		18	CRC/GC/API-II	PHILIPPINES
	No	Yes (18) *	* 18	17	18	CRC/GC/API-II	POLAND
	No	Yes (4/12) *	* 18	* 17	18 see note * 21	CRC/GC/API-II	PORTUGAL
	No	No			?	CRC/GC/API-II	QATAR
	No	Yes (12/18) *	* 20	* 18	18	CRC/GC/API-II	ROMANIA
* 17 * 11	Yes *	Yes (18)	18		18	CRC/GC/API-II	RUSSIAN FEDERATION
  not stated * 15	Yes	No		* 18	18	CRC/GC/API-II	RWANDA
	No	Yes *	* 18		18	CRC/GC/API-II	SAO TOME & PRINCIPE
	No	No *			?	CRC/GC/API	SAUDI ARABIA
	Yes *	Selective (24)	?	18	18	CRC/GC/API-II	SENEGAL
	No	Yes (24)	?		18	CRC/GC/API-II	SEYCHELLES

A	B	C	D	E	F
Country	Population (millions)	Armed Forces Paramilitaries and/or Militias Armed Opposition Groups	Child Soldiers Total Number	As % of force	Girls?(%)
SIERRA LEONE	4.5	* 8,000	1,000–2,000	*16–24	Yes (3)*
RUF		* 5,000	3,000–4,000	* 60–80	
miscellaneous *		5,000	?000	?	
SINGAPORE	2.9	70,000			
SLOVAKIA	5.3	* 43,000			
SLOVENIA	2.0	9,550			
SOMALIA	7.1	* >17,000	?000	?	
SOUTH AFRICA	40.7	* 79,440	* 12	<1	
paramilitary police		138,000			
SDUs		* >10,000	?000	?	
SPAIN	40.0	197,500			
paramilitary		* 75,000			
SRI LANKA	17.6	112,000–117,000			
paramilitary forces		110,000			
LTTE		19,000	?000	?	Yes(15) *
SUDAN	25.0	79,700	16,000	20	No
People's Defence Forces		* 15,000	5,000	20	Maybe
SPLA		50,000	* 10,000	20	Yes
SURINAME	0.4	1,800			
SWAZILAND	0.8	see note			
SWEDEN	8.6	53,350			
SWITZERLAND	6.9	* 400,000			
SYRIA	13.0	320,000			
gendarmerie		* 8,000			
TAIWAN	20.9	376,000	?0000	?	
paramilitary forces		26,500			
TAJIKISTAN	5.6	7,000–9,000	?00	?	No report
Border Guards		1,200			
UTO		5,000			
TANZANIA	25.1	34,600			
militias		* 100,000			
THAILAND	57.8	266,000			
TOGO	3.8	6,950			
paramilitary		* 800			
TONGA	0.1	* 300			
TRINIDAD & TOBAGO	1.2	2,100			
TUNISIA	8.5	35,000			
paramilitary forces		* 13,500			
TURKEY	60.0	639,000			
paramilitary forces		182,000			
Village Guards (militias)		70,000			
PKK		* 15,000	* 4,000	* 25–30	Yes (>10)*
TURKMENISTAN	3.9	16,000–18,000			

V

Lowest Age Recorded	Armed Conflict?	Conscription and Duration of Service (in months)	Legal Recruitment Ages		Voting Age	Treaties CRC GC API APII	
			Conscripts	Volunteers			
* 5 7	Yes	No			21	CRC/GC/API-II	SIERRA LEONE
	No	Yes (24/36)	18		21	CRC/GC/API-II	SINGAPORE
	No	Yes (12) *	* 18		18	CRC/GC/API-II	SLOVAKIA
	No	Yes (7) *	* 18		18	CRC/GC/API-II	SLOVENIA
11	Yes	See note			?	GC	SOMALIA
* 11	Yes *	No *		* 17	18	CRC/GC/API-II	SOUTH AFRICA
	No	Yes (9) *	* 19		18	CRC/GC/API-II	SPAIN
8	Yes	No		18	18	CRC/GC	SRI LANKA
10 7 9	Yes	Yes (36) *	*18 (16?)	* 16	21	CRC/GC	SUDAN
	No	No			18	CRC/GC/API-II	SURINAME
	No	No *			18	CRC/GC/API-II	SWAZILAND
	No	Yes(7/20) *	* 18		18	CRC/GC/API-II	SWEDEN
	No	Yes (3.5) *	20		18	CRC/GC/API-II	SWITZERLAND
	No	Yes (30)	19		18	CRC/GC/API-II	SYRIA
	No	Yes (24)	* 15		-	-	TAIWAN
16	Yes	Yes *	* 18		18	CRC/GC/API-II	TAJIKISTAN
	No	Yes (24)	18		18	CRC/GC/API-II	TANZANIA
	No	No *			18	CRC/GC	THAILAND
	No	Selective (24)			18	CRC/GC/API-II	TOGO
	No	No			21	CRC/GC	TONGA
	No	No *		* 18	18	CRC/GC	TRINIDAD & TOBAGO
	No	Yes	20	18	20	CRC/GC/API-II	TUNISIA
* 7	Yes	Yes (18) *	* 20		18	CRC/GC	TURKEY
	No	Yes (18)			18	CRC/GC/API-II	TURKMENISTAN

A	B	C	D	E	F
Country	Population (millions)	Armed Forces Paramilitaries and/or Militias Armed Opposition Groups	Child Soldiers		
			Total Number	As % of force	Girls?(%)
UGANDA	16.6	40,000–55,000			
paramilitary forces		1,000			
LRA		* 5,000–7,000	* 3,000–5,000	* 67	Yes (31) *
WNBF		1,000–2,000	?00	?	
other opposition groups		> 2,000	?00	?	
UKRAINE	52.3	387,400			
UNITED ARAB EMIRATES	1.9	64,500			
UNITED KINGDOM	58.1	* 209,400	* 5,528	* 2	Yes (11) *
IRA, UDF, etc.		not known	?00	?	
UNITED STATES OF AMERICA	256.6	1,447,600	* 6,700	* <1	Yes (14) *
Civil Air Patrol		51,000			
URUGUAY	3.2	25,600			
UZBEKISTAN	21.0	65,000–70,000			
VANUATU	0.2	* 300			
VENEZUELA	20.7	57,000			
National Guard		* 22,000			
VIETNAM	70.9	* 492,000			
YEMEN	12.5	66,300			
paramilitary		* 20,000			
YUGOSLAVIA	10.6	114,200	?00	?	
ZAMBIA	7.8	21,600			
ZIMBABWE	10.7	39,000			

G Lowest Age Recorded	H Armed Conflict?	I Conscription and Duration of Service (in months)	J Conscripts	K Volunteers	L Voting Age	M Treaties CRC GC API APII	
* 5 11	Yes	Selective *	* 18	* 18	18	CRC/GC/API-II	UGANDA
	No	Yes (12/24) *	* 18		18	CRC/GC/API-II	UKRAINE
	No	No			?	CRC/GC/API-II	UNITED ARAB EMIRATES
* 13	Yes *	No		* 16		CRC/GC/API-II	UNITED KINGDOM
	No	No	* 18	17	18	GC	UNITED STATES OF AMERICA
	No	No *		* 18	18	CRC/GC/API-II	URUGUAY
	No	Yes (18)	?		18	CRC/GC/API-II	UZBEKISTAN
	No	No			18	CRC/GC/API-II	VANUATU
	No	Selective (24)	18		18	CRC/GC	VENEZUELA
	No	Yes (24/36) *	18		18	CRC/GC/API-II	VIETNAM
	No	Yes (24/36)	?		18	CRC/GC/API-II	YEMEN
	No *	Yes (12) *	* 18	17	18	CRC/GC/API-II	YUGOSLAVIA
	No	No			18	CRC/GC/API-II	ZAMBIA
	No	No *			18	CRC/GC/API-II	ZIMBABWE

(Reference letters relate to the columns of the table.)

AFGHANISTAN

D & E The figure of 45% comes from the case study and relates to all principal factions active in 1993, before the Taliban forces became a significant element on the scene. The total given here is calculated from this proportion. Reports increasingly speak of a large – but unquantified – proportion of the Taliban fighters being teenagers; the total suggested here is a very round guess.

G The Case Study does not refer to any participants aged lower than 13.

I & J In 1978, according to the Case Study, the conscription age was 22 years, but this was progressively lowered as the civil war continued. By the early 1990's, a conscription age of 15 was being reported, but there is no evidence that the legal age limit was ever lowered below 18, whatever the *de facto* position in the face of widespread illegal forced recruitment. At the time of writing it is clear that compulsory recruitment continues to be practised by all factions in the areas which they control, but the legislative basis, if any, is unknown.

ALBANIA

G Widespread civil disturbances in the Spring of 1997 led to the capture of military arms supplies by groups in which boys down to the age of 10 were conspicuous, and the emergence of various anti-government groupings. It is possible that with the restoration of order the armed opposition groups and/or the participation of minors have ceased to exist.

H Defined as a low-intensity conflict by the PIOOM Databank, Leiden University, the Netherlands, who adopt a broader definition than other conflict-monitoring institutions.

I & J   Source: War Resisters' International (see "Sources" on p. 205)

ANGOLA

D & E   The total numbers of child soldiers quoted are early estimates
of the number to be demobilised on both sides after the signing of the Lusaka Accords in November 1994, and the percentages have been derived from these numbers. Although under
the terms of the accord there have been subsequent well-pub-
licised demobilisations of minors, it is by no means clear how
far either this process, or the parallel integration of the remaining Government and UNITA forces into a new national army,
have progressed, nor that all fresh recruitment of minors has
ceased.

ANTIGUA & BARBUDA

I   Source: Information supplied by War Resisters' International
CONCODOC project.

ARGENTINA

I & J   Source: UN Document E/CN.4/97/99 (see "Sources" on p. 205).
"Law No.24.429, promulgated on 5 January 1995, establishes a
voluntary military service yet reserves for Congress the right to
conscript 18-year-olds for a period of service not exceeding
one year. Such conscription may be ordered, when for specified
reasons, an inadequate number of volunteers present themselves for military service." (p.15)

ARMENIA

H   Armenian forces were involved in a war with Azerbaijan over
the Armenian-populated enclave of Nagorno-Karabakh.
Although that conflict continues at a lower intensity, Nagorno-
Karabakh having unilaterally declared its independence,
Armenia itself is no longer directly involved.

AUSTRALIA

D & F   Actual recruitment of minors in 1997. (Communication from Australian Department of Defence, 5/1/98.)

I & J   Ibid. "The minimum age for conscription is 18 but Australia is not conscripting at the present time".

K   The Australian Defence Force "undertakes not to deploy 17 year olds into hostilities until they reach the age of 18". (Communication from Department of Defence, 5/1/1998.)

AUSTRIA

C & D   Communication from the Permanent Mission of Austria to the United Nations, Geneva, 9/12/1997.

I   Source: War Resisters' International (See "Sources" on p. 205).

AZERBAIJAN

H   Defined as a low-intensity conflict by the PIOOM Databank, Leiden University, the Netherlands, who adopt a broader definition than other conflict-monitoring institutions.

I & J   A form of conscription is imposed by the self-proclaimed "Armenian Republic of Nagorno-Karabakh.

BANGLADESH

H   Defined as a low-intensity conflict by the PIOOM Databank, Leiden University, the Netherlands, who adopt a broader definition than other conflict-monitoring institutions.

K   Communication from the Permanent Mission of Bangladesh to the United Nations, Geneva, 18/11/97.

BENIN

I   Source: The Guinness World Fact Book (see "Sources" on p. 205)

BHUTAN

B   Government figures tend to be only some 60% of those given by independent sources. The Government contests the size of the population of Nepali origin.

C Source for all figures: Case Study. Since 1990, the army and the police have allegedly doubled in numbers and the militia groups have been formed. For the army total, a statement to the National Assembly by the Chief of Army in August 1995 is quoted. The total strength of the militias is not known; the figure of 2,200 represents only those detachments specifically mentioned in the case study.

E The Case Study quotes testimonies that one detachment of 200 soldiers contained three very young soldiers "(not more than 15 years"), and that another of 400 contained 15 who were under 16 years. It also quotes a testimony that in one group of 300 militia recruits there were 25 to 30 "very young boys". If typical, these observations would imply the approximate percentages given here, from which may be extrapolated speculative totals. Given the ages mentioned and the unknown total size of the militias, these estimates may well be conservative.

F According the the Case Study only the Royal Bhutan Police takes female recruits, and no claim is made of child recruitment into that force.

G Case Study.

BOLIVIA

I & J Source: UN Document E/CN.4/97/99 (see "Sources" on p. 205)

BOSNIA & HERZEGOVINA

B As of 1991. A substantial proportion of the population fled during the war, and many have yet to return.

D According to the Case Study, between 3,000 and 4,000 children participated in the 1991–1995 war in the former territory of Yugoslavia, the vast majority in Bosnia and Croatia. Barnen och vi, 6 (1995), published by Rädda Barnen, quotes the enrolment of children as young as 11 in the regular forces, while the Case Study quotes one instance of a child of 10 taking part in the fighting.

H   Defined as a continuing low-intensity conflict by the PIOOM Databank, Leiden University, the Netherlands, who adopt a broader definition than other conflict-monitoring institutions.

I & J   Source: War Resisters' International (see "Sources" on p. 205) "During the war military service could last indefinitely." The current period of service is unknown. The autonomous authorities in Republika Srpska – the Bosnian Serb Republic – operate conscription to their own forces, the current strength of which is unknown. The period of service was reduced in December 1996 from eighteen to nine months, and the current age of liability is believed to be 18. However, the Case Study indicates that during the war recruits were officially accepted during the year in which they would turn 17, and that in practice many volunteered earlier.

BULGARIA

C   Source for militias: The Guinness World Fact Book (see "sources" on p. 205)

I & J   Source: War Resisters' International (see "Sources" on p. 205)

BURKINA FASO

C   Figure for armed forces includes gendarmerie.

BURMA/MYANMAR

E & G   Source: Case study. The proportions of minors in Government armed forces are the highest and lowest proportions cited for any particular regiment in the interviews carried out for that study.

BURUNDI

D   The Rädda Barnen database quotes a UN Report of 29/10/97

that the Burundian armed forces "had been recruiting an estimated 5,000 to 7,000 extra men and training some 2,000 high
school graduates, which would bring their number up to nearly 30,000 troops". The implication is that the high school graduates are members of the armed forces: in the light of the legal
recruitment age and the absence of any more specific information it seems reasonable to assume that the number of minors
in the armed forces will not be less than the number of recruits
involved in this training programme.

G   Case study for Burundi.

CAMBODIA

C   The figure for the strength of the armed forces includes per-
        haps 35,000 in local militias. Between July 1997 and February
        1998 there was renewed conflict between Funcinpec forces,
        numbering at least 2,000, who supported the deposed joint
        Prime Minister, Prince Rannaridh, and those parts of the armed
        forces who continued to uphold the Hun Sen government.
        However, as both sides were drawn from the pre-existing armed
        forces, and no detailed information about their composition is
        available, they are not differentiated in the table. At the time of
        writing, negotiations are under way with a view to reintegrat-
        ing the two forces. As for the Khmer Rouge, the case study
        quoted a strength of 12,000, but there have subsequently been
        substantial defections.

E & G   Case Study.

I & J   Source: UN Document E/CN.4/97/99 (see "Sources" on p. 205)

CAMEROON

H   Border conflict with Nigeria.

CANADA

D   Of 3,545 recruits into the armed forces in 1997, 204 were aged
        17 and 24 aged 16. (Communication from Canadian Mission to

the United Nations, Geneva, 16/2/98.) The figure given in the table assumes that the number of 16-year-olds is relatively constant from year to year.

CAPE VERDE

I   Source: UN Document E/CN.4/97/99 (see "Sources" on p. 205)

CENTRAL AFRICAN REPUBLIC

I   Source: The Guinness World Fact Book (see "Sources" on p. 205)

CHAD

I   Source: The Guinness World Fact Book (see "Sources" on p. 205)

CHINA

C   In the first edition of this book a total of 12 million paramilitaries was quoted, but it has not been possible to confirm this figure.

H   The confrontation between the security forces and members of the Uighur ethnic group in Xinkiang is defined as a low-intensity conflict by the PIOOM Databank, Leiden University, the Netherlands, who adopt a broader definition than other conflict-monitoring institutions.

COLOMBIA

paramilitaries: These are independent of Government control although in many cases there is evidence of close co-operation between them and the armed forces against the opposition guerrilla groups.

D   Defensoria del Pueblo [Ombudsman's Office], Victimas de la violencia: El conflicto armado en Colombia y los menores de edad (Boletin No.2 Sistema de Seguimento y Vigilancia – La Ninez y sus Derechos), Santafe de Bogata 1996. The figure for the army is the number of minors recruited in 1995 and the first part of 1996, according to the national army recruitment office. These represented 22% recruits to the army. Independent sources would suggest a higher figure. The same report indicated that 63% of conscripts were recruited into the police, on which basis the number of minors involved might be expected to be at least as high as for the army.

E   Source as above. The CNG (Guerrilla co-ordination council) admits to 7–10% child soldiers. Sources in the armed forces and former guerrillas themselves quote total figures of around 30% – up to 80% in urban guerrilla units and lower in rural areas.

G   The lowest age for guerrillas was quoted in the Case Study, that for paramilitaries in the report from the Defensoria del Pueblo cited above.

J & K   Under Act No. 418 of December 1997 the conscription of persons under the age of 18 was prohibited, but the recruitment of volunteers over 16 years of age was allowed, subject to parental consent, although such persons might not be used in armed confrontations or sent to combat zones. Previously students had been liable to recruitment on leaving high scool.

COMOROS

C   Source: The Guinness World Fact Book (see "Sources" on p. 205)

CONGO, DEMOCRATIC REPUBLIC

C & D   Since the first edition of this book went to press, an armed opposition group, the ADFL, has not only emerged in what was Zaire, but has overthrown the government, renamed the country the Democratic Republic of Congo, and has itself been challenged by an armed uprising. It is clear that this has been one of the areas of the world in which child soldiers have been most active in the last three years, but with the composition and alignment of the various forces changing so rapidly figures are hard to pin down. There is, for example, very incomplete information on the present distribution of the armed and paramilitary forces of the former Zaire, who totalled over 70,000. The number of demobilised ADFL child soldiers comes from the United Nations Office for the Co-ordination of Humanitarian Affairs Integrated Regional Information Network (IRIN Update No. 346 for Central and Eastern Africa, 3/2/98).

G   In July 1998 President Kabila made a widely-publicised appeal to male citizens between 12 and 20 to enlist in order to "defend the country". The Economist of 5/9/98 reports, "Although the government denies enlisting child soldiers, uniformed boys as young as 12 were visible at every roadblock, resting their chins on the end of their gun barrels."

I, J, L   In the absence of any information on legislation passed since the establishment of the DRC, the situation which pertained in Zaire is quoted.

COSTA RICA

J   Constitutional provision exists for the establishment of conscripted armed forces, should that become necessary.

COTE D'IVOIRE

C & I   Source: The Guinness World Fact Book (see "Sources" on p. 205). The Rädda Barnen database quotes a total of 13,900 including paramilitaries.

CROATIA

H   Defined as a continuing low-intensity conflict by the PIOOM Databank, Leiden University, the Netherlands, who adopt a broader definition than other conflict-monitoring institutions.

I & J   Source: War Resisters' International (see "Sources" on p. 205). Although all men are liable for conscription from the age of 16 (women, too, may be liable under certain unspecified circumstances), registration does not take place until the year of the 18th birthday, and the military service itself usually takes place at the age of 18. Croatia has a specific provision to exempt from military service those who have joined as volunteers.

CUBA

C: Conscripts also work on the land; presumably the "Youth Labour Army", whose strength is estimated as 65,000.

I & J   Source: UN Document E/CN.4/97/99 (see "Sources" on p. 205)

CYPRUS

C   Excluding mainland Turkish forces in the occupied zone, which number about 30,000.

I & J   Source: War Resisters' International (See "Sources" on p. 205). The second set of figures relate to the self-styled "Turkish Republic of Northern Cyprus".

CZECH REPUBLIC

C,I,J   Communication from the Permanent Mission of the Czech Republic to the United Nations, Geneva, 13/1/98. War Resisters' International (see "Sources" on p. 205) give the formal age of liability as 19, but confirm that recruitment takes place at 18.

DENMARK

C   Communication from the Danish Ministry of the Interior, 27/1/98.

D & E   Actual number of recruits aged 17 in 1997, of whom four (11%) were women. (Source as above.)

I   War Resisters' International (see "Sources" on p. 205) report that as there is a surplus of potential conscripts, recruitment is by lot, and, in fact, "[m]ost conscripts in the armed services have volunteered to serve (in 1993 ... 82.4%)". (p. 31).

J & K   Although the conscription age is 18, there was provision enabling potential conscripts to apply to perform their military service early if this was more convenient, eg. with regard to the timing of their higher education. In exceptional circumstances, volunteers could also be admitted as professional soldiers at the age of 17. In 1996, 40 young men opted to commence their obligatory military service at the age of 17; all 38 17-year-old recruits in 1997 were however volunteers. (Communication from Ministry of Interior referred to above.) In June 1998, the Prime Minister announced that hencefoth all recruitment of under-18s would cease.

DJIBOUTI

B   Including over 100,000 refugees, principally from Somalia.

C   The armed forces total may include about 2000 paramilitary and the forces of the former FRUD opposition.

DOMINICAN REPUBLIC

I   A legal basis for conscription exists, but it is not currently enforced. (Information supplied by War Resisters' International CONCODOC project.)

EGYPT

G   "A 'teenage boy' was among 39 males brought to trial for membership of the al-Gama'a al-Islamiya [armed opposition group] in November 1997.

H   Defined as a low-intensity conflict by the PIOOM Databank, Leiden University, the Netherlands, who adopt a broader definition than other conflict-monitoring institutions.

EL SALVADOR

I   A new law on military service adopted in 1992 gives a legislative basis for conscription, but has not been put into effect.

EQUATORIAL GUINEA

C   Source for paramilitaries: The Guinness World Fact Book (see "Sources" on p. 205)

ERITREA

I & J   Source: Rädda Barnen.

ESTONIA

I, J, K   UN Document E/CN.4/97/99, (see "Sources" on p. 205), quoting European Council of Conscripts Organisations.

ETHIOPIA

Opposition groups: Information on opposition armed groups is based on figures quoted in the Case Study, regarding the number and age distribution of OLF captives taken after the first major confrontation between the partners in the alliance which had defeated the Derg government in 1992. The OLF is the largest of several armed opposition groups known to be currently active in different parts of the country, but on whom more up-to-date information is scanty.

C & D   The case study provides evidence of the participation of minors in the civil war which culminated in 1992 with the victory of the EPRDF forces and their allies (including those who seceded to form the new state of Eritrea) over the Derg regime. Although the national armed forces are of a similar size to the former EPRDF army, there has been a major demobilisation programme, including possibly more than 20,000 child soldiers, and a parallel recruitment aimed in part at redressing the ethnic imbalance in the forces. The case study expressed the hope that with the civil war over this campaign would be unlikely to impinge on children, but there are allegations that it has, which are the basis for the entries in this table.

E   Whereas during the civil war the then government forces were exclusively male, between a quarter and a third of the strength of the EPRDF was female.

H   The PIOOM Databank, Leiden University, the Netherlands, who adopt a broader definition than other conflict-monitoring institutions, identified five ongoing low-intensity conflicts in Ethiopia before the outbreak of a border conflict with Eritrea in June 1998.

FINLAND

I & J   Source: War Resisters' International (see "Sources" on p. 205)

FRANCE

C   Source for gendarmerie: The Guinness World Fact Book (see "Sources" on p. 205)

I & J   Source: War Resisters' International (see "Sources" on p. 205). Conscription is to be abolished within the next few years, being replaced by compulsory one-day courses for all sixteen-year-olds.

L   The voting ages refer to the National Assambly and the Senate respectively.

GABON

C   Source for paramilitaries: The Guinness World Fact Book (see "Sources" on p. 205)

GAMBIA

C   Source for armed police force: The Guinness World Fact Book (see "Sources" on p. 205)

K   Communication from the Department of State for Defence, Republic of the Gambia, 7/1/98.

GERMANY

I & J   Source: War Resisters' International (see "Sources" on p. 205)

K   The source for the Bundesgrenzenschutz is a communication received from the Permanent Mission of the Federal Republic of Germany to the Office of the United Nations and the other International Organisations, Geneva (1/12/97). According to the Federal Ministry of Defence less than 0,5% of recruits enter at 17 (April 1998).

GREECE

I & J   Source: War Resisters International (see "Sources" on p. 205)

GUATEMALA

PAC Militias and URNG: Figures in the table relate to the situation during the civil war. (The PAC militias were formally abolished in 1994.)

D   Source: Case Study. During the months of May and June 1995, the Attorney General's Office for Human Rights (Procuradoria de Derechos Humanos) received 596 complaints from young men who claimed to have been recruited by force, and obtained the release of 333 of them, of whom 148 were under 18 (presumably at the time of release). If indeed these cases referred only to a two-month period, they could have reflected anything between 1,000 and 4,000 forced under-age recruitments per annum.

G   Case Study.

H   The civil war formally ended with the peace accord of 29 December 1996.

I   The "Global Human Rights Accord" of 1994 stipulated that pending the enactment of a new law on military service such service should be voluntary. (Information supplied by War Resisters' International CONCODOC project.)

GUINEA

C   Source for paramilitaries: The Guinness World Fact Book (see "Sources" on p. 205)

GUINEA-BISSAU

H   Armed conflict broke out in June 1998, but at the time of preparing the table no details of the opposing forces were available.

I & J   Source: UN Document E/CN.4/97/99 (see "Sources" on p. 205)

GUYANA

C   Source for paramilitaries: The Guinness World Fact Book (see "Sources" on p. 205)

I   Source: UN Document E/CN.4/97/99 (see "Sources" on p. 205)

HAITI

H   Defined as a low-intensity conflict by the PIOOM Databank, Leiden University, the Netherlands, who adopt a broader definition than other conflict-monitoring institutions.

I   Source: UN Document E/CN.4/97/99 (see "Sources" on p. 205)

HONDURAS

Opposition Groups: Honduras has at no time in the recent past been classified as involved in an internal armed conflict, under any definition. Even so, the case study documents the existence of opposition guerrilla groups which recruit minors.

D   The Case Study quotes estimates from the annual report of CODEH (The Honduran Committee for Human Rights) that there were some 3,690 cases of forced recruitment, mainly of minors, in the five-year period 1988–1992.

I   "Decree No. 24–94 was passed in May 1994 establishing a voluntary military service during peacetime. The amendment reserves for the Congress the right to conscript." (UN Document E/CN.4/97/99 (see "Sources" on p. 205), p18, quoting evidence submitted by the National Interreligious Service Board for Conscientious Objectors).

HUNGARY

C,I,J,K Communication from the Permanent Mission of the Republic of Hungary to the United Nations, Geneva, 9/12/1997, which also points out that volunteers are accepted only after completion of the compulsory military service, so that "the age ... is 19 in practice".

INDIA

H   The Military Balance 1996/7 (International Institute of Strategic Studies, London), reports that the Indian army is supporting the police and paramilitary forces in their struggle with twelve armed separatist groups,

I, J, K   Statement by Indian representative during meeting in January 1997 of UN Working Group on an Optional Protocol to the Convention on the Rights of the Child. Legal provision exists for conscription, but it is not currently in force.

INDONESIA

H   Government forces have been involved in armed conflict with separatist groups in Irian Jaya (the Indonesian half of New Guinea), in the Aceh region of Sumatra, and in East Timor, annexed by Indonesia in 1975. In addition, the end of 1997 and the beginning of 1998 saw more widespread civil disturbances.

IRAN

Basij: The Basij, or Popular Mobilisation Army numbered anything up to a million volunteers during the Iran/Iraq war, and it was to this force that the reports of very young soldiers related. It continues to provide a volunteer reserve in time of peace, albeit with much smaller numbers, relying heavily upon youths in its recruitment, and (see report in Guardian Weekly, 3/5/98) functions as a "moral police".

KDP   See also under Iraq.

I & J   Source: UN Document E/CN.4/97/99 (see "Sources" on p. 205)

IRAQ

KDP   See also under Iran.

C   The figure for the PUK includes 22,000 "tribesmen"; the figure
        for the KDP 25,000 "tribesmen".

I   Source: UN Document E/CN.4/97/99 (see "Sources" on p. 205)

K   Statement by Iraqi representative during meeting in January
        1997 of UN Working Group on an Optional Protocol to the
        Convention on the Rights of the Child. A programme known as
        "Saddam's Youth" involves preparatory weapons training for 10
        to 15 year-olds, but these are enrolled only for the duration of
        the training and there is no evidence that they are considered
        as members of the armed forces.

IRELAND

I & K   Source: War Resisters' International (see "Sources" on p. 205)

ISRAEL

C: The figure quoted for the PNLA (the military wing of the PLO)
        includes a considerable proportion based in other countries of
        the region.

I & J   Source: UN Document E/CN.4/97/99 (see "Sources" on p. 205)
        "Men and women over 18 [liable], although non-Druze Israeli
        Arabs and Druze women are exempt.". The shorter period of
        service is for female conscripts.

ITALY

I   Source: War Resisters' International (see "Sources" on p. 205)

J & K   Communication from the Permanent Mission of Italy to the
        United Nations, Geneva, 23/1/98.

L   The voting ages refer to the Chamber of Deputies and the
        Senate, respectively.

JAPAN

I & K   Communication from the Permanent Mission of Japan to the
        United Nations, Geneva, 1998 (not dated). "The only exception
        [to the age-limit of 18] is some 15 and 16 years old are recruit-

ed as Self-Defence Forces youth cadets to be trained to become technical specialists ... During the first three years of their four year training period, the youth cadets follow the equivalent curriculum as other senior high school students, in addition to learning basic matters necessary for Self-Defence Forces Personnel."

JORDAN

D   Communication from the Permanent Mission of the Hashemite Kingdom of Jordan to the United Nations, Geneva, 23/6/98.

I   According to UN Document E/CN.4/97/99 (see "Sources" on p. 205), "Jordanian Conscript Service was suspended indefinitely in 1992 and all members of the armed forces are regular volunteers". This was not however mentioned by the Jordanian Mission in the communication cited against D above.

I,J,K   Communication from Jordanian Mission cited against D above. Volunteers spend one year in training centres before deployment and are not therefore available for deployment or participation in hostilities before the age of 18.

KOREA, NORTH

I & J   Information supplied by War Resisters' International CONCO-DOC project.

KOREA, SOUTH

I, J, K Communication from the Permanent Mission of the Republic of Korea to the United Nations, Geneva, 14/1/98.

KYRGYSTAN

I   Rädda Barnen's sources report that conscription does exist, but give no details. Against this, it should be noted that the strength of the armed forces has fallen by more than 40% since Kyrgystan became independent at the breakup of the Soviet Union.

LAOS

H   Defined as a low-intensity conflict by the PIOOM Databank, Leiden University, the Netherlands, who adopt a broader definition than other conflict-monitoring institutions.

LATVIA

C   Including National Guard. (Communication from Latvian Ministry of Defence, 25/11/97).

I,J,K   Source: UN Document E/CN.4/97/99 (see "Sources" on p. 205). "The military service itself is performed between the ages of 18 and 25."

LEBANON

A   The Case Study documents also the participation of children in the various factional militias during the 1975–90 civil wars, specifically quoting testimonies from former child soldiers held on criminal charges. (In December 1994, five years after the supposed end of hostilities, 646 children were in detention on charges involving arms).

F   The Rädda Barnen database quotes estimates that there are between 30 and 50 girls in the ranks of Hizbollah and between 20 and 25 in the South Lebanon Army. The Case Study documents without numbers the participation of girls in Palestinian fringe groups.

G   With the exception of Hizbollah, the minimum ages are from testimonies reported in the case study.

LIBERIA

C to G Details from the Case Study, quoting the First Progress Report of the Technical Committee of the Task Force for Disarmament, Demobilisation and Reintegration, August 1995.

I   At the time of writing there is no effective authority in a position to enforce conscription.

LIBYA

J & K   State Party Report of the Libyan Arab Jamahiriya to the United Nations Committee on the Rights of the Child, 26/9/96, Document CRC/C/28/Add.6.

LITHUANIA:

C   Source for National Guard: The Guinness World Fact Book (see "Sources" on p. 205)

I, J, K   Source: UN Document E/CN.4/97/99 (see "Sources" on p. 205). It is possible to opt to perform the military service at the age of 18.

MACEDONIA

I & J   Source: War Resisters' International (see "Sources" on p. 205)

MADAGASCAR

I   "Including conscription for civil purposes." Source: The Military Balance 1996/97 (Institute for Strategic Studies, London).

MALI

C   Source for paramilitaries: The Guinness World Fact Book (see "Sources" on p. 205)

I   Source: The Military Balance 1996/97 (Institute for Strategic Studies, London).

MAURITANIA

I   Source: The Military Balance 1996/7, quoted in the Rädda Barnen database.

MAURITIUS

K   "The minimum age at which one could apply to join the Special Mobile Force (classified as paramilitary in your book) would not be below 18 years." (Communication from Mauritius Mission to the United Nations, Geneva, 26/1/98.)

MEXICO

H   The situation in the province of Chiapas is defined as a low-intensity conflict by the PIOOM Databank, Leiden University, the Netherlands, who adopt a broader definition than other conflict-monitoring institutions.

I   The one year period of compulsory military service consists of a weekly Saturday morning drill. (Information supplied by War Resisters' International CONCODOC project.)

J & K   Written statement by Mexico to meeting in January 1997 of UN Working Group on an Optional Protocol to the Convention on the Rights of the Child.

MOLDOVA

I   Source: UN Document E/CN.4/97/99 (see "Sources" on p. 205)

MOROCCO

C   Source for paramilitaries: The Guinness World Fact Book (see "Sources" on p. 205)

I   It has been alleged that the Polisario forces fighting for the independence of the Western Sahara enforce "conscription" at the age of 17 on their client population.

MOZAMBIQUE

H   Defined as a continuing low-intensity conflict by the PIOOM Databank, Leiden University, the Netherlands, who adopt a broader definition than other conflict-monitoring institutions.

NEPAL

H   Defined as a low-intensity conflict by the PIOOM Databank, Leiden University, the Netherlands, who adopt a broader definition than other conflict-monitoring institutions.

NETHERLANDS

C   Including about 3,600 paramilitary Royal Military Police (Koninklijke Marechaussee), who recruit on the same system as the other services.

D,E,F   Figures derived from information on the recruitment of minors during the period 1992–1994 provided by the Permanent Mission of the Kingdom of the Netherlands to the United Nations, Geneva (communication dated 14/6/1996). It is possible that the situation may have subsequently changed as a result of the ending of conscription. (see below)

I & J   Source: War Resisters' International (see "Sources" on p. 205) "... in 1992 parliament decided to suspend the call-ups. On 29 February 1996 the last conscripts were called up to perform six months' service ... The Netherlands armed forces are now completely professional. With the passing of a new law on military service [March 1997] ... conscription still exists, but there are no regulations on the performance of military service. This means registration for the draft is still taking place: all 17-year-old men are to go on a military register. But the recruits are no longer summoned for a medical examination and there is no military call-up. However, at any time government may introduce regulations on the length of military service and the Minister of Defence can issue call-up notices." (p. 63)

K   Raised from 16 to 17 in 1997. (Statement by Netherlands representative during meeting in January 1997 of UN Working Group on an Optional Protocol to the Convention on the Rights of the Child.)

NEW ZEALAND

D,E,F   Proportion of minors and of girls supplied by the Permanent Mission of New Zealand to the United Nations, Geneva (communication dated 12/7/96), and total number calculated by reference to armed forces total.

J   As of 1996, some 10–15% of naval recruits were aged under seventeen; whereas under the Defence Act 1990 no person in the Army or Airforce was liable for active service outside New Zealand while under the age of 18 years, the limit in the navy was sixteen years and six months (communication of 12/7/96 quoted above). It is believed that all recruitment under the age of 17 has now ceased; the Ministry of Foreign Affairs and Trade (communication dated 18/3/98) states categorically " ... the minimum age of legal recruitment into the New Zealand Armed Forces is 17 years old and the minimum age for deployment to active service is 18 years.".

NORWAY

C   Strength of Home Guard from Guinness World Fact Book (see "Sources" on p. 205)

D,E,F, Communication from Permanent Mission of Norway to the United Nations, Geneva, 10/2/98. (Percentages derived from figures quoted.)

I   Source: War Resisters' International (see "Sources" on p. 205)

J & K   Communication from Norwegian Mission cited above.

PAKISTAN

K   Communication from the Permanent Mission of Pakistan to the United Nations, Geneva, 16/12/97. "17–22 years for officers and 16–25 years for enlisted men."

PANAMA

I   Source: UN Document E/CN.4/97/99 (see "Sources" on p. 205). No details given on recruitment age or length of service.

PAPUA NEW GUINEA

H   On the island of Bougainville. A cease fire agreement was signed in October 1997.

PARAGUAY

D,E,F,G Case study for Paraguay.

PERU

A   Sendero Luminoso = Shining Path.

C   Shining Path's strength was considerably depleted by the secur-
        ity forces' 1992 offensive. At their peak they are believed to
        have forcibly recruited several thousand children from indige-
        nous communities in areas under their control.

G   With the exception of MRTA (for which it quoted a minimum
        age of 12), the source for minimum ages is the case study for
        Peru.

J   Source: Case Study. Men and women liable.

K   Source: Case Study. Men only, and only with the father's writ-
        ten authority.

PHILIPPINES

NPA   The source for information on child soldiers in the NPA is the
        case study for the Philippines. The reported proportions, from
        which the estimated total has been calculated, varied between
        units.

POLAND

I & J   Source: War Resisters' International (see "Sources" on p. 205)

PORTUGAL

C   Source for paramilitaries: The Guinness World Fact Book (see
        "Sources" on p. 205)

I   Source: War Resisters' International (see "Sources" on p. 205)

J & K   Communication from the Permanent Mission of Portugal to the
        United Nations, Geneva, 17/2/98. For admission into the
        National Republican Guard there is no minimum age, but the

recruits must first have completed their military service.

QATAR

C & K   Communication from the Permanent Mission of the State of Qatar to the Office of the United Nations, Geneva, 18/6/98. Although 18 is stated to be the minimum recruitment age, reference is made to a recruitment age of "17 years for Qatari Citizen and 18 years other citizens".

ROMANIA

I   Source: War Resisters' International (see "Sources" on p. 205)

J & K   Communication from the Permanent Mission of Romania to the United Nations, Geneva, 14/1/98. This states further that registration for military service takes place during the year the young man reaches the age of 19, or at the end of "non-obligatory" secondary education, but that the service itself is not performed until the age of 20. Legal provision exists for conscription in time of war during the year in which young men reach the age of 18, but they would not go into active service until after a period of training.

RUSSIAN FEDERATION

E & G   Case study for Chechnya. Although the existence of under-age recruitment into Government armed forces is reported, such cases are seen as isolated exceptions rather than a systematic practice.

H   In Chechnya. The situation in the neighbouring republic of Daghestan is also defined as a low-intensity conflict by the PIOOM Databank, Leiden University, the Netherlands, who adopt a broader definition than other conflict-monitoring institutions.

RWANDA

A   Figures for population and armed forces in Rwanda are as at the eve of the massacres and refugee outflow of April 1994.

C   The current strength of the Government (ie. former RPF) forces is 62,000, and that of the opposition is believed to be about 6,000, including the rump of the former Government forces and militias, probably including children of both sexes, and dispersed over Rwanda itself, the Democratic Republic of Congo, Burundi and Tanzania

D, E, G Case study for Rwanda. "The number of child soldiers per se in Rwanda's former military was negligible. Nonetheless... thousands of boys between 15–18 years old were part of the local militia. (Also) unknown numbers of boys and girls between 6–18 years old were involved (as) informants within the context of communal purges."

SAO TOME E PRINCIPE

C   Source: The Guinness World Fact Book (see "Sources" on p. 205)

I & J   Source: Decree Law No. 3/83 of 4 March.

SAUDI ARABIA

C   Source for paramilitaries: The Guinness World Fact Book (see "Sources" on p. 205)

I   Source: UN Document E/CN.4/97/99 (see "Sources" on p. 205)

SENEGAL

H   The separatist struggle in the Casamance region is defined as a low-intensity conflict by the PIOOM Databank, Leiden University, the Netherlands, who adopt a broader definition than other conflict-monitoring institutions.

SEYCHELLES

C   Source for paramilitaries: The Guinness World Fact Book (see "Sources" on p. 205)

SIERRA LEONE

"miscellaneous" This category includes various "self-defence" militias, most importantly what are described as "Kamajor traditional Hunters" who resisted the military coup of May 1997.

C   The integration of the forces formerly belonging to the opposition RUF into the Government armed forces was interrupted by the military coup of May 1997, and the outbreak of a new phase of civil war, which has continued beyond the reversal of that coup. The figures given here are estimates dating from the end of the previous phase of civil war.

D, E, F, G   Figures relating to the erstwhile Government forces are taken from a case study reported in McCallin, M., The reintegration of young ex-combatants into civilian life, ILO, 1995. Sources relating to the RUF include "The Children's War: Towards Peace in Sierra Leone" (The Women's Commission for Refugee Women and Children, 1997) which states, "As many as 80% of rebel soldiers are between the ages of seven and fourteen".

SLOVAKIA

C   Communication from the Permanent Mission of the Slovak Republic to the United Nations, Geneva, 19/11/97.

I   Source: War Resisters' International (see "Sources" on p. 205)

J   Communication from the Slovak Mission cited above.

SLOVENIA

I & J   Source: War Resisters' International (see "Sources" on p. 205)

SOMALIA

C & I   There is no effective government. The figures given relate to the various factions currently operating in Somalia, including the self-styled "Republic of Somaliland" in the former British protectorate.

SOUTH AFRICA

C   A communication dated 15/1/98 from the Permanent Mission
        of South Africa to the United Nations, Geneva, gives the
        strength of the National Defence Force as 76,000 plus 51,000
        reserves. The figure for SDUs is from the Case Study.

D   As at 18/10/96. (Address to the General Assembly Debate on
        Item 106 on Promotion and Protection of the Rights of
        Children by the Permanent Representative of South Africa to
        the United Nations, New York, 11/11/96).

G   Case study for South Africa.

K   Communication from the South African Mission cited above.
        "The South African National Defence Force accepts volunteers
        at the age of 17 but restricts the minimum eligible age for com-
        bat duty to 18 years."

SPAIN

C   Source for paramilitaries: The Guinness World Fact Book (see
        "Sources" on p. 205)

I & J   Source: War Resisters' International (see "Sources" on p. 205).
        Liability is from 18, but conscripts are not called up until the
        age of 19. The government has indicated that it intends to end
        conscription by the year 2003.

SRI LANKA

F   Based on the LTTE's own figure of 3,000 female members and
        the assumption that the age distribution is similar to that of
        males. Participation of girls down to the age of 13 has been
        reported.

SUDAN

C   The figure of 15,000 for the People's Defence Forces is based on
        a narrow definition of the active strength. The recruit force is
        much larger; villages and residential areas, students and civil

servants may be incorporated wholesale for short periods with an obligation to continue an affiliation.

D   Based on the proportions quoted in Rädda Barnen's database. It is, however, worth noting that a figure of 13,000 child soldiers was attributed to the leader of a faction which split from the SPLA in a press article in 1991. ("Des enfants entre famine et enrolement", by Stephen Smith and Patricia Antinian, Liberation, Paris 21/9/91.)

I, J, K   Source: The Military Balance 1996/7, quoted in the Rädda Barnen database. The legal minimum recruitment age was 18. It is not clear whether this provision has been formally repealed, but a major recruitment campaign was launched by the Government in Autumn 1997 which was explicitly aimed at youth aged 16 and upwards. The Minister of State at the Ministry of Defence was quoted as warning parents not to try to hide their children, implying that the recruitment concerned would be compulsory.

SWAZILAND

C & I   In 1990, according to The Guinness World Fact Book (see "Sources" on p. 205), the armed forces totalled 2,700 and there was also a paramilitary police force. "Compulsory military service" lasted for two years but was "not universal". The information provided by the government to the Commission on Human Rights (UN Document E/CN.4/1997/99, see "Sources" on p. 205) implies that this has now been abolished, but no more recent information on the armed forces is available.

I & J   Source: War Resisters' International (see "Sources" on p. 205).

SWITZERLAND

C & I   Communication from the Permanent Mission of Switzerland to the International Organisations, Geneva, 15/1/98. A very large proportion of the total armed strength consists of conscripts on short training courses. The Military Balance gives the "active total" of full-time servicemen as only 3,300.

SYRIA

C   Source for gendarmerie: The Guinness World Fact Book (see "Sources" on p. 205)

TAIWAN

J   Source: European Council of Conscripts' Organisations, ECCO Echo, No.3, Nov. 1996. In the face of public disquiet, the authorities admitted at the end of 1995 that over the past five years 2,355 young conscripts had died in the course of their military service. (L'Etat du Monde 1997, Editions Decouverte, Paris, p.486). Although the source did not specify this, it is likely from the conscription age that all or nearly all of these were under 18.

TAJIKISTAN

I & J   Information from Rädda Barnen database.

TANZANIA

C   Source for militias: The Guinness World Fact Book (see "Sources" on p. 205)

THAILAND

I   Conscription was abolished in October 1997.

TOGO

C   Source for paramilitaries: The Guinness World Fact Book (see "Sources" on p. 205)

TONGA

C   Source: The Guinness World Fact Book (see "Sources" on p. 205)

TRINIDAD & TOBAGO

I & K   Communication received from the Permanent Mission of the Republic of Trinidad and Tobago to the United Nations, Geneva, 13/2/98.

TUNISIA

C   Source for paramilitaries: The Guinness World Fact Book (see "Sources" on p. 205)

TURKEY

PKK All information from the Case Study. (The number of child soldiers is estimated from the other figures quoted). The PKK itself is the source for the total strength in 1994, which they hoped to double. The Military Balance 1996/7, quoted in the Rädda Barnen database, gives a figure of 5,000, plus 50,000 "support militia".

I & J   Source: War Resisters' International (see "Sources" on p. 205) The Case Study quotes a conscription age of 18, but nevertheless reports no evidence of any recruitment of minors into government armed forces.

UGANDA

C, D, E It is known that the LRA "recruits" almost exclusively by abducting children. The Military Balance 1996/97 quotes the strength of the group as 2,000 (half based over the Sudanese border.) However, it is estimated that 3,000 to 5,000 abductees are currently held by the group. The simplest way to reconcile these figures is to assume that the former relates to the long-standing membership, to which the number of abductees should be added. In view of this uncertainty, there can be no precise estimate of the overall proportion of children within the group, but approximately two thirds of abductees who escape are aged under 18, and it is known that the youngest abductees rarely escape.

I   Rädda Barnen database, quoting Human Rights Watch.

J & K   Communication from the Permanent Mission of Uganda to the United Nations, Geneva, dated 28/1/98.

UKRAINE

I & J   Source: UN Document E/CN.4/97/99 (see "Sources" on p. 205)

UNITED KINGDOM

D, E, F As of 1/7/98. Source: Defence Analytical Services Agency.

H   In Northern Ireland. Following the 1997/98 "peace process" there are hopes that the period of armed conflict may now have finally ended.

UNITED STATES OF AMERICA

D & E The exact percentage is 0,46; the total given here is calculated from that figure.

F   Overall proportion of women in armed forces.

I   Provision exists for selective call-up in time of war.

URUGUAY

I & K   Communication from Permanent Mission of Uruguay to the United Nations, Geneva, dated 8/12/97.

VANUATU

C   Source: The Guinness World Fact Book (see "Sources" on p. 205)

VENEZUELA

C   Source for Nationel Guard: The Guiness World Fact Book (See "Sources" on p. 205).

VIETNAM

C   In the late 1980's the armed forces were estimated at over one and a quarter million; they have subsequently been drastically reduced.

I   Lengths of service provided by War Resisters' International CONCODOC project.

YEMEN

C   Source for paramilitaries: The Guinness World Fact Book (see "Sources" on p. 205)

YUGOSLAVIA

H   Armed conflict broke out in the Kosovo region in May/June 1998. At the time of preparing the table no details could be obtained of the opposition Kosovo Liberation Army.

I & J   Source: War Resisters' International (see "Sources" on p. 205)

ZIMBABWE

I   Source: UN Document E/CN.4/97/99 (see "Sources" on p. 205)

ANNEX 2

Issues and Questions for the Case Studies

1. **Context:** Describe the conflict, including the type of conflict, background, duration, different stages or phases, and how it ended (if it has), in order to set the scene for your answers and to facilitate a general understanding of the situation. An overview of other relevant information would also be helpful, e.g. the percentage of children in the population, the difficulties you have encountered in doing the case study, and the nature of your (or your organisation's) work in the country.

2. **Recruitment:** This section seeks to identify how many and which children become involved in armed forces, how they become involved, what functions they perform initially and whether this changes over time. Particular attention should be paid to the affect of gender in relation to all these issues.

2.1 Numbers in armed forces, with a breakdown of ages (lowest age, under 15, 15–18 years) and sex. How does this relate to the number of adults in the armed forces? At what age are children considered to be adults for this purpose? Is this different for girls? Is age the criterion used or is there another definition, e.g. height? Is there a minimum age for legal recruitment (compulsory or voluntary) under national law? Are those responsible for recruitment aware of the international standards about the minimum age for recruitment?

2.2 Types of activity performed. Does this vary according to age, sex, method of recruitment, type of armed forces, duration of service? Are girls forced/expected/encouraged to render sexual services? Does this vary according to their age and the other functions which they perform?

2.3 Type of armed forces in which children are serving, e.g. governmental, opposition, self-defence forces. Are there armed forces which have no child soldiers? Were children involved in any peacetime armed forces?

2.4 Method of recruitment: compulsory, forced, voluntary, induced? If voluntary, what were the reasons for volunteering? Were benefits offered to encourage volunteers (financial, other)? Are there situations where the armed groups are considered to be better off than the governmental armed forces? Are some categories of children more vulnerable to recruitment than others? Which categories, e.g. poor, less-educated, those from families with problems, children separated from their parents/family (give reasons for separation), certain age group(s), boys/girls, rural/urban, refugees and/or displaced, those in conflict zones, specific ethnic/indigenous or religious groups? To what type of recruitment are they vulnerable and why? Are particular categories targeted for recruitment and if so, which ones, why and by whom? Are certain categories not targeted for recruitment and if so, why?

3. Treatment in the armed forces: This section seeks to identify how children are treated or affected while they are in the armed forces. The similarities and differences between children and adults both in treatment and in the effects of their experience should be addressed. Particular attention should be paid to the effect of age and gender in relation to all these issues.

3.1 Are families involved in the armed forces in any way, e.g. cooking for their soldier family members, caring for wounded family members?

3.2 Conditions of work: are the children paid and how does this compare with adult soldiers? Does it vary according to age, gender, type or duration of service? Do they get leave, or time off for recreation or visits to their family (do adult soldiers)? Are they given training? Does this include training in any skills which might be useful in civilian life? Are they subjected to toughening-up procedures? What punishments are given for disobedience? Do they receive any education? Do they get preferential feeding or other treatment in comparison with adult soldiers? Are they subject to abusive treatment, such as provision of drugs, alcohol, the rendering of sexual services, other?

3.3 Duration of involvement in armed forces: how long do children stay in the armed forces? What affects the duration of their service? Do the tasks they perform and the treatment they receive vary with the duration of service?

3.4 Numbers killed and injured, and types of injury: how many child soldiers (including those in support functions) get killed? How does this compare with the numbers of adult soldiers killed and with the number of child civilians killed? Do the deaths result from different causes? How many child soldiers are injured? What types of injury do they suffer? How do these compare with the numbers and types of injury to adult soldiers and to child civilians? Are there injuries which are child specific, e.g. hearing loss caused by firing automatic weapons? How do age and gender affect numbers and types of death and injury of child soldiers?

3.5 Provision of treatment for injured: Are injured child soldiers provided with medical care? Are they returned to their families for care? If so, what happens to those without families? Are they abandoned to their own devices? How does their care compare with the provision for injured adult soldiers?

3.6 Numbers and reasons for demobilisation other than at the end of an armed conflict: e.g. invalidity, capture, other? Are there circumstances in which organised demobilisation has taken place other than at the end of the conflict?

3.7 Treatment on capture: How are child soldiers treated when captured? Are they treated as prisoners of war? Is any provision made for access (by families or others)? Is any special provision made for them as children, e.g. education? Are they returned to their families? Are they recruited into the capturing force? Does their treatment differ between different parties to the conflict? Does age, sex, involvement in killings/atrocities, method of recruitment, injuries or other factors affect the way in which they are treated?

QUESTIONNAIRE B

The purpose of this questionnaire is to supplement Questionnaire A by undertaking an in-depth analysis, including observations and suggestions (taking account of gender) of the reasons for the involvement of children as active participants in the conflict and what happens to them afterwards.

4. Reasons for involvement: This section is intended to identify the factors which decide why children become involved in some conflicts and not in others, or change from passive observers/victims into active participants. It should include an analysis of methods of recruitment and involvement, including the reasons for volunteering or circumstances of engagement, and information on any preventive strategies employed to reduce or eliminate recruitment of children – whether by those in authority or by communities, families, the children themselves or others – and their effectiveness. Observations, suggestions and recommendations in relation to actions and strategies to decrease or prevent such involvement in the future should be included.

4.1 Have there been different stages in the conflict? How has this affected the involvement of children? At what stage in the conflict did children first become active participants or have they been involved throughout? Historically, have children always been involved in armed conflicts (if so, from what age) or is this something new or different in size or kind?

4.2 How did children become involved? This is an opportunity to add anything to the answers already provided in Questionnaire A.

4.3 What specific factors led to this involvement: e.g. shortage of adult soldiers, the attitude of the military/political leadership, ideology, vulnerability of certain groups of children, the need (or perceived need) to support or protect the family, peer group pressure? Was it a policy decision by those in military/political authority? Do families/communities support the involvement of their children, if so why, e.g. material benefits, ideology/religion/culture?

4.4 What factors inhibited or prevented such involvement, e.g. availability of education, ideology, stable family environment, opposition by families (mothers?) or communities, opposition by military/political/religious or other leaders?

4.5 Were any specific strategies employed to reduce/prevent child recruitment, including voluntarily, if so what, by whom and how effective were they, e.g. keeping schools open, sending children away from the area or out of the country, reuniting separated children with families, making financial or material contributions instead, provision of alternative activities for children (e.g. as first aid workers)?

4.6 How do the military view the involvement of children in their armed forces and their relationship to them? How do they view the involvement of children in opposing armed forces?

5. **The post-service situation**. The purpose of this section is to understand the short-, medium- and long-term effects of children's participation in conflict. In this regard, when answering the following questions it would be most helpful if information describing the situation of former child soldiers could be compared where appropriate with that of former adult soldiers and of children who were not actively involved in the conflict.

5.1 Education: What proportion of former child soldiers are attending school? Is their present level of education appropriate to their age? Are special programmes considered necessary to meet their educational needs? Why, and what form do/will they take? Who will be responsible for their implementation – e.g. government/military; local/international organisations? In addition to enabling the children to regain the education they lost as a result of their involvement in the conflict, are educational activities also considered to assist in a more general way in re-educating the children to civilian life? How is this accomplished? Do teachers identify any problems working with child soldiers? Are teachers given support and assistance in their work? Is it considered that they will require any special training to work with former child soldiers? What is the attitude of the former child soldiers to education?

5.2 Physical and mental well-being: Do former child soldiers have any special health problems – e.g. physical disabilities; sexually transmitted diseases? Are the children considered to behave differently from other children who were not actively involved in the conflict? How do people describe their behaviour? Are these issues specific to certain groups/categories of child soldiers, e.g. girls/adolescents/children from a particular family/ethnic background? Do these health/social issues create difficulties in their daily lives and affect their successful reintegration? What forms of intervention could help in overcoming these difficulties? Is there any indication that former child soldiers are engaged in criminal/violent activities? In situations where some time has passed since their involvement in the conflict, are there indications of long-term social/emotional consequences for the children, e.g. family discord, difficulties in assuming their appropriate roles in society?

5.3 Employment/Vocational Training: Do former child soldiers experience difficulties in finding employment? What factors influence this situation? Where children have come from rural homes, do they usually resume their family's traditional means of livelihood? Are vocational training schemes available for former child soldiers? Are they appropriate to local circumstances – availability of necessary materials, opportunities for employment etc? Are the child soldiers integrated with other children on these training schemes? Are other forms of job training available, e.g. apprenticeship to local craftspeople/assistance to learn the family's traditional means of income generation? Who initiates and is responsible for these schemes? Do personnel on these schemes identify any special problems working with former child soldiers? Is special training/support available to them in their work with the children?

5.4 Community Attitudes and Involvement: Do the children's families and communities identify any social/economic consequences due to their participation in the conflict? Do they affect their reintegration into civil society? How do they cope with these problems? Are some problems specific to certain groups of children – e.g. age/gender-related/family circumstances/knowledge of role played/acts committed by the child during conflict? Is there any community mobilisation to assist the children's reintegration? Who initiated this, parents/teachers/community leaders...? What measures do the community think could be implemented to facilitate the children's social reintegration? Did the families/community support the recruitment of children and, if so, why? Has their attitude changed towards this form of recruitment and, if so, what brought this about? Is there support for recruitment of children only from certain sectors in the community – who, why, and how do other community members cope with this? Do the families/community identify any benefits to themselves or their children as a result of participation in conflict? What are they?

5.5 The Children: Every effort should be made to include in the case study discussions with child soldiers themselves. These discussions should reflect their attitudes and concerns on the above issues 5.1 to 5.4 in order that we can incorporate their views about the factors that contribute to

264

their reintegration into civil society and the difficulties they are experiencing. Where girls have been involved in the conflict, their special concerns should be noted.

6. Planning for Reintegration.

6.1 For countries where conflict has ended. Did the parties to the conflict sign a peace agreement? If so, was the participation of children as soldiers recognised/specified in any way in the peace agreement? Did this recognition reflect concern for the effects of militarisation on children *per se*, and the consequent need to assist them to reintegrate into civil society? If so, what is considered to have influenced the decision to incorporate recognition of this particular group of soldiers in the peace agreement? Have former child soldiers received any demobilisation benefits? What form do they take? Are they comparable to those awarded to adult soldiers? Have programmes for rehabilitation/reintegration of former child soldiers been implemented? What form do they take and who is responsible? What factors are considered (a) to assist this process and (b) to hinder programme development? Are any special groups of former child soldiers targeted as in special need of programmes? Who are they and how are they identified? Are any children excluded from the programme? Is the military supportive of/involved in the children's rehabilitation? Are the children monitored in any way? Who is responsible for the monitoring? What issues are prioritised?

6.2 For countries where conflict is on-going. Have children actively serving as soldiers been officially demobilised or otherwise ceased participating in the conflict? Who are these children – members of government forces/rebel groups/militias/other? Did these children receive any benefits – demobilisation pay/clothing/tools, etc. to assist them on their return to their families/communities? Is their progress monitored in any way? What factors influenced the end of their participation – a specific policy decision/resistance from families and community/pressure from local/international organisations/changes in the nature of the conflict? Is there any planning now for the future rehabilitation/reintegration of children presently involved in the conflict? Who initiated this planning process and which organisations (governmental/local/ international) are involved? What factors influenced the decision to develop such programmes? What factors are hindering the planning process? What are considered to be the priority issues for the children in addressing rehabilitation/reintegration? Are the children involved in the planning in any way?

Contributors

Jesuit Refugee Service, Asia-Pacific
International Catholic Child Bureau, Asia
Connell Foley, Cambodia
Colectivo por la Objecion de Conciencia al Servicio Militar, Colombia
Lutheran World Federation, El Salvador
Yitayew Alemayehu, Shoangizaw Chane, Ethiopia
Ismet G. Imset, London, Great Britain
Defensa de la Ninez Internacional, Guatemala
El Comite de Paz y Justicia de la Iglesia Menonita Hondurena y Participaciónes,
  Honduras
Defence for Children International, Israel
Georges J. Assaf, Lebanon
Children's Assistance Program, Inc, Liberia
Ministerio da Coordenacao de Accao Social, Mozambique
Centre for the Victims of Torture, Nepal
World Vision International, Nicaragua
Mine Clearance Planning Agency, Pakistan
Servicio Paz y Justicia, Paraguay
Rädda Barnen, Peru
Centre for Antiwar Action, Belgrade, Serbia
National Institute for Public Interest Law and Research, South Africa
Images Asia, Thailand
World Vision International, Uganda
Office of the UN High Commissioner for Refugees

ANNEX 4

List of Abbreviations

ABSDF	All-Burma Student's Democratic Front (Burma/Myanmar)
ADFL	Allied Democratic Forces for the Liberation of Congo/Zaire
AIS	Armee Islamique de Salut (Algeria)
ANC	African National Congress (South Africa)
API	Additional Protocol I of 1977 to the Geneva Conventions of 1949
APII	Additional Protocol II of 1977 to the Geneva Conventions of 1949
BRA	Bougainville Revolutionary Army (Papua New Guinea)
CRC	Convention on the Rights of the Child of 1989
ECOMOG	West African States' Peacekeeping Mission (Liberia)
ELN	Ejercito de Liberación Nacional (National Liberation Army; Colombia)
EPRDF	Ethiopian People's Revolutionary Democratic Front
EPL	Ejército Popular de Liberación (Popular Liberation Army; Colombia)
EZLN	Ejercito Zapatista Nacional de Liberation (Mexico)
FAES	Fuerzas Armadas El Salvador (Salvadorean Armed Forces)
FARC	Fuerzas Armadas Revolucionarias Colombianas (Revolutionary Armed Forces of Colombia)
FIS	Front Islamique du Salut (Islamic Salvation Front; Algeria)
FMLN	Farabundo Martí para la Liberación Nacional (Farabundo Martí Liberation Movement (El Salvador)
FRELIMO	Frente de Libertaçao de Moçambique (Front for the Liberation of Mozambique)
FRETILIN	Frente Revolucionaria de Timor Leste (Revolutionary Front of East Timor)
FRUD	Front for the Restoration of Unity and Democracy (Djibouti)
GC	Geneva Conventions I–IV of 1949
GIA	Groupe Islamique Armé (Armed Islamic Group; Algeria)

ICRC	International Committee of the Red Cross
IRA	Irish Republican Army (Northern Ireland)
JSS/SB	Jana Sanghati Samiti/Shanti Bahini (Bangladesh)
KA	Karenni Army (Burma/Myanmar)
KDP	Kurdish Democratic Party
KNU/KNLA	Karen National Union/Karen National Liberation Army (Burma/Myanmar)
LRA	Lord's Resistance Army (Uganda)
LTTE	Liberation Tigers of the Tamil Eelam (Sri Lanka)
MILF	Moro Islamic Liberation Front (Philippines)
MNLF	Moro National Liberation Front (Philippines)
MRTA	Movimiento Revolucionario Tupac Amaru (Peru)
MTA	Mong Tai Army (Burma/Myanmar)
NAPARAMAS	A Military-religious movement wich operated in northern Mozambique during the war
NGO	Non-governmental organisation
NMSP/MNLA	National Mon State Party/Mon National Liberation Army (Burma/Myanmar)
NPA	New People's Army (Philippines)
NPFL	National Patriotic Front of Liberia
OAU	Organisation of African Unity
OLF	Oromo Liberation Front (Ethiopia)
PAC	Pan African Congress (South Africa)
PAC	Patrullas de Autodefensa Civil (Civil Defence Patrols; Guatemala)
PDF	People's Defence Force (Sudan)
PDF	People's Democratic Front (Burma/Myanmar)
PDK	Parti Democratique de Kampuchea; Cambodia ("Khmer Rouge")
PLO	Palestine Liberation Organisation
PKK	Kurdistan Worker's Party (Turkey)
PNLA	Palestinian National Liberation Army
PUK	Patriotic Union of Kurdistan (Iraq)
RBG	Royal Body Guard (Bhutan)
RBP	Royal Bhutan Police (Bhutan)
RENAMO	Resistencia Nacional Mozambique (Mozambique National Resistance)
RPF	Rwandan Popular Front

RUF	Revolutionary United Front (Sierra Leone)
SDU	(unofficial) Self-Defence Unit allied to ANC (South Africa)
SLA	South Lebanon Army
SPLA	Sudan People's Liberation Army
SPU	(unofficial) Self-Protection Unit allied to Inkatha (South Africa)
UDF	Ulster Defence Force (Northern Ireland)
ULIMO	United Liberation Movement (Liberia)
ULNLF	United Lao National Liberation Front
UNHCR	United Nations High Commissioner for Refugees
UNICEF	United Nations Children's Fund
UNITA	Uniao Nacional para a Independencia Total de Angola
UTO	United Tadjik Opposition
URNG	Unidad Revolucionaria Nacional Guatemalteca (Guatemalean National Revolutionary Unit; Guatemala)
UWSP/UWSA	United Wa State Party/United Wa State Army (Burma/Myanmar)
WNBF	West Nile Bank Front (Uganda)

Rädda Barnen's Project on Child Soldiers

Rädda Barnen – Save the Children Sweden – has for many years campaigned against the use of child soldiers. This issue is not only connected to Article 38 of the Convention of the Rights of the Child, but actually cuts across several of the core principles of the Convention. Children in armed conflict is also one of the basic target groups to which Rädda Barnen's programme activities are devoted. In 1995, the organisation launched a new project on child soldiers in line with its previous efforts.

The project has four components:

1. A newsletter "Children of War" which is produced in English and Spanish and is published four times a year. The primary purpose of the newsletter is to arouse public opinion and stimulate action against the use of children as soldiers.

2. An information database on child soldiers (ChildWar), containing numbers, ages, sex, relevant forces, etc broken down by country and by armed forces. This is available on the Internet through Rädda Barnen's home page www.rb.se.

3. A documentation database (Soldoc) with references to materials on child soldiers. It contains references that relate to documents with information on field experience and best practices, the militarisa-

tion of children and peacetime military use of children. The Soldoc database is also available on the Internet.

4. An established contact database including over 2,000 addresses worldwide of organisations and individuals concerned with the issue of child soldiers.

As of 1998 this project forms Rädda Barnen's Global Information Centre on Child Soldiers. Through the Information Centre Rädda Barnen provides journalists, researchers, organisations and governmental representatives, with information packages, publications, booklets, footage, etc.

International Coalition to Stop the Use of Child Soldiers

During 1998 Rädda Barnen's advocacy efforts on child soldiers have intensified by means of an international campaign to address the plight of children used as soldiers or at risk of recruitment. The main objective of the campaign is to stop the use of child soldiers by improving national and international legal standards prohibiting the military recruitment of children and use of child soldiers under 18 years of age. Rädda Barnen considers and promotes international experiences and lessons learnt in preventing the recruitment of children and the demobilisation, rehabilitation and social reintegration of child soldiers.

Rädda Barnen also take an active part in the steering committee of the international Coalition to Stop the Use of Child Soldiers (for more information see Annex 6). The Coalition's primary objectives are the adoption of, and adherence to, an Optional Protocol to the (UN) Convention on the Rights of the Child (CRC) prohibiting the military recruitment and use in hostilities of any person younger then 18 years of age (the straight 18's position); and the recognition and enforcement of these standards by all armed forces and groups, both governmental and non-governmental.

Regional support and documentation of best practices

One of Rädda Barnen's main objectives is to document best practices of our child soldier programmes in order to improve future implementation. We therefore work in close co-operation with our programme offices to develop tools of best practices on prevention, protection, demobilisation, reintegration and rehabilitation. Field visits and follow-up studies are important aspects in parallel with national and local advocacy efforts.[424]

Training of peacekeeping personnel

Rädda Barnen has since 1993 been training Swedish and African international peace-keepers in the rights of the child and the importance of children's rights to protection. Approximately 6,000–7,000 peace-keepers have participated in the courses provided jointly by the Swedish Armed forces and Rädda Barnen. Rädda Barnen has intensified its training efforts by means of a child soldiers training component. We have focused on how to deal with child soldiers when they are caught, surrender, etc. This component has also been presented in a peace keeping seminar in Addis Abeba during spring 1998 and will be practised in Nordic Peace operation in September 1998. The Child Soldier component has generated great interest but is still in the process of development.

Building Partnership

In order to improve the impact of our advocacy work with children in armed conflict Rädda Barnen seek the participation and collaboration of key actors. Building partnership and networks with governmental representatives, UN agencies and other NGOs is important.

UNICEF, UNHCR, UNHCHR, ICRC and the Special Representative of the Secretary General for Children in Armed Conflict are important partners. So are regional organisations such as the OAU and the EU and NGOs, including Amnesty International, Human Rights Watch, the Quakers, the International Federation Terres des Hommes, Jesuit Refugee Service, OXFAM, ANPPCAN, Handicap International, DCI, LWF, Vietnam

---

[424] See for example the ARC-module, "Action for the Rights of Children – A Training and Capacity-building Initiative on Behalf of Refugee Children and Adolescents, 1998, London (International Save the Children Alliance & UNHCR). The ARC modules can be requested from: International Save the Children Alliance, 275-281 King Street, London W6 9LZ, United Kingdom, or at the UNCHR, Case Postale 2500, CH-1211 Geneva 2 Depôt, Switzerland.

Veterans, the World Council of Churches, Peace and Justice Commissions in Latin America and national/local NGOs, etc.

For more information contact:

Henrik Häggström
Project Manager/Editor
Rädda Barnen
107 88, Stockholm
Sweden
Tel: +46 8 698 90 86, Fax: +46 8 698 90 12
e-mail: henrik.haggstrom@rb.se
Web address: http://www.rb.se

or
Irene Opira
Assistant
Rädda Barnen
107 88, Stockholm
Sweden
Tel: +46 8 698 91 26, Fax: +46 8 698 90 13
e-mail: irene.opira@rb.se

ANNEX 6

The Coalition to Stop the Use of Child Soldiers

A newly-formed international coalition of non-governmental organisations was launched in June 1998 to tackle the appalling phenomenon of child soldiers. More than 300,000 children under 18 years of age are believed to be currently participating in armed conflicts around the world. Hundreds of thousands more are members of armed forces or groups and could be sent into combat at almost any moment.

Although most child soldiers are between 15 and 18 years of age, significant recruitment starts at 10 – the age at which a child can handle an AK-47 – and the use of even younger soldiers has been recorded. The risks to children and the impact on them of involvement in armed conflict, as well as the consequent dangers for the civilian population as a whole, demand that they neither be recruited nor allowed to serve as soldiers in any armed forces.

The number of children being recruited or volunteering to fight in current conflicts is constantly growing, despite the protection provided to children by existing international law, which prohibit the use of children under fifteen years. In addition, it is widely agreed that the existing international minimum age of 15 for recruitment into the armed forces and participation in hostilities is too low and should be raised to 18. The Working Group of the UN Commission on Human Rights mandated to draft an Optional Protocol to the Convention on the Rights of the Child on the involvement of children in armed conflict has so far failed to agree on a text which would prohibit the recruitment and participation of all children.

The Coalition to Stop the Use of Child Soldiers has been formed with the primary objectives of promoting the adoption of, and adherence to, an Optional Protocol to the (UN) Convention on the Rights of the Child prohibiting the military recruitment and use in hostilities of any person younger than 18 years of age; and the recognition and enforcement of this standard by all armed forces and groups, both governmental and non-governmental. The Coalition is led by a Steering Committee of seven international non-governmental organizations (NGOs) – Amnesty International, Defence for Children International, Human Rights Watch, International Federation Terre des Hommes, International Save the Children Alliance (represented by Save the Children Sweden – Rädda Barnen), the Jesuit Refugee Service and the Quaker United Nations Office, Geneva. A small Secretariat, comprising a full-time Coordinator and a part-time information officer, has been set up in Geneva, Switzerland.

There is a pressing need to inform and mobilise international public opinion against all forms of recruitment - both voluntary recruitment and conscription - and all forms of participation of under-18s and thereby to create the necessary political will in favour of strict standards prohibiting the use of children as soldiers. The Coalition will therefore adopt a high-profile media and lobbying campaign designed to raise awareness of the extent of the problem and to promote action to tackle the problem, including the adoption of an effective Optional Protocol. The support of concerned governments, international and non-governmental organisations as well key international figures able to influence public opinion will be sought.

For further information please contact the International Secretariat,
Tel: ++ 41 22 917 8169; Fax: ++ 41 22 917 8082
E-mail: child.soldiers@wanadoo.fr
Website: http://www. child-soldiers.org

Bibliography

For further information on the literature listed below, including the 26 case studies produced for the book, please contact Rädda Barnen's Project on Child Soldiers (see address in Annex 5). The case studies are listed on pages 16–17.

Amnesty International, *Sierra Leone: Human rights abuses in a war against civilians.* London, Sept. 13, 1995.

Amnesty International, *Sierra Leone: Prisoners of war?: Children detained in barracks and prison.* London, Aug. 12, 1993.

"Angola: Civilians devastated by fifteen year war." In: *News From Africa Watch.* 1991 (vol. 3), no. 1.

Balian, H., *Armed conflict in Chechnya, its impact on children.* Covcas Center for Law and Conflict Resolution, Arlington, Va., 1995. (Report prepared for the UN study on the Impact of Armed Conflict on Children).

Blomqvist, U., "Social work in refugee emergencies: capacity building and social mobilisation: the Rwanda experience". In: McCallin, M. (ed.), *The psychological well-being of refugee children: research, practice and policy issues* (second edition), International Catholic Child Bureau, Geneva, 1996.

Bonnerjea, L., *Family tracing: a good practice guide.* Save the Children, London, 1994. (Save the Children Development Manual no. 3).

Boothby, N., Upton, P. & Sultan, A. "Boy soldiers of Mozambique." In: *Refugee Participation Network.* 1992: no. 12 (March).

"Boy soldiers: a new and ruthless breed of warriors." In: *Newsweek,* 1995 (vol. CXXVI), no. 6.

Children, genocide and justice: Rwandan perspective on culpability and punishment for children convicted of crimes associated with genocide. Save the Children Federation USA, Kigali, 1996.

Children in war: community strategies for healing. Save the Children USA, University of Zimbabwe and Duke University, Durham, N.C., 1995.

Defensoria del Pueblo (Human Rights Ombudsman), *Victimas de la Violencia: El conflicto Armado en Colombia y los Menores de Edad* (Victims of Violence: The Armed Conflict in Colombia and the Minors), Bogota, Colombia, May 1996.

Dütli, M. T., "Captured child combatants." In: *International Review of the Red Cross,* 1990: no. 278 (September–October).

Fozzard, S., *Surviving violence: a recovery programme for children and families.* International Catholic Child Bureau, Geneva, 1995.

Gibbs, G., "Post-war social reconstruction in Mozambique: reframing children's experience of trauma and healing". In: *Disasters,* 1994 (vol. 18), no. 3.

Goodwin-Gill, G. & Cohn, I., *Child soldiers: the role of children in armed conflicts.* Clarendon Press, Oxford, 1994.

Hampson, F.J., *Legal protection afforded to children under international humanitarian law.* University of Essex, Colchester, 1996. (Report prepared for the UN study on the Impact of Armed conflict on Children).

Human Rights Watch/Africa, Arms Project, *Angola: Arms trade and violations of the laws of war since the 1992 elections.* New York, 1994.

Human Rights Watch/Africa, Children's Rights Project, *Children of Sudan: slaves, street children and child soldiers.* New York, 1995.

Human Rights Watch/Africa, Children's Rights Project, *Easy prey: child soldiers in Liberia.* New York, 1994.

Images Asia, *"No childhood at all":* a report about child soldiers in Burma. – Rev. ed. – Chiangmai, Thailand, Images Asia, *1997.*

International Save the Children Alliance & UNHCR: *"Action for the Rights of the Children – A Training and Capacity-Building Initiative on Behalf of Refugee Children and Adolescents,* 1998, London.

Jareg, E., *Rehabilitation of child soldiers in Mozambique.* Oslo: Redd Barna, 1993. (Note prepared for the Secretary General of Redd Barna).

Jareg, E. & Jareg, P., *Reaching children through dialogue.* Basingstoke: Macmillan Press, 1994.

Jareg, E. & McCallin, M., *The rehabilitation of former child soldiers: report on a training workshop for caregivers of demobilised soldiers: Freetown, Sierra Leone, September 1st to 3rd 1993.* Geneva: International Catholic Child Bureau, 1993.

Jongman, A.J. & Schmid, A.P., "World conflict map, 1994–1995." In: *PIOOM Newsletter and Progress Report.* 1995 (vol. 7), no. 1.

Laffin, J., *"Boys in Battle"* Abelad-Schuman (1966) Ltd, 8 King Street WC 2.

Maramba, P., *Reintegration of demobilized combatants into social and economic life: the Zimbabwean experience.* Geneva, 1995. (Report on a consultancy for the ILO).

McCallin, M., *The reintegration of young ex-combatants into civilian life: expert meeting on the design of guidelines for training and employment of ex-combatants, Harare, 11–14 July 1995.* Geneva: International Labour Office, 1995.

Philipson, D., *"Band of Brothers – Boy Seamen in the Royal Navy 1800–1956.* Sutting Publishing Limited (1996), Great Britain, Cornwall.

Price-Cohen, C., "Considerations affecting the implementation of the United Nations Convention on the Rights of the Child in situations of forced migration." In: M. McCallin (ed.), *The psychological well-being of refugee children: research, practice and policy issues* (second edition). Geneva: International Catholic Child Bureau, 1996.

Reichenberg, D. & Friedman, S., "Traumatized children: healing the invisible wounds of children in war: a rights approach." In: Danieli, Y., Rodley, N. S. & Weisaeth, L. (eds.), *International responses to traumatic stress.* Baywood Publishing Company, Inc., Amityville, N.Y., 1996.

Ressler, E., Tortorici, J. M. & Marcelino, A., *Children in war: a guide to the provision of services.* New York: UNICEF, 1993.

Roberts, S. & Williams, J., *After the guns fall silent.* Vietnam Veterans of America Foundation, Washington D.C., 1995.

Save the Children Alliance, *Promoting psychosocial well-being among children affected by armed conflict and displacement: principles and approaches.* Geneva, 1996. (Paper prepared as a contribution to the UN study on the Impact of Armed conflict on Children).

Tolfree, D., *Restoring playfulness: different approaches to assisting children who are psychologically affected by war or displacement.* Stockholm: Rädda Barnen, 1996.

Tortorici, J.M., "Peace education in Nicaragua." In: M. McCallin (ed.), *The psychological well-being of refugee children: research, practice and policy issues* (second edition). Geneva: International Catholic Child Bureau, 1996.

United Nations, *Comments on the report of the Working Group on a draft optional protocol to the Convention on the Rights of the Child on involvement of children in armed conflicts.* Geneva, 1995. (UN document E/CN.4/1996/WG.13/2, 23 Nov.)

Further reading drawn from the Rädda Barnen database

For further information on the literature listed below, please contact Rädda Barnen's Documentations Centre on Child Soldiers (see address in Annex 5)

Afghanistan: the war against children: a summary report. Eureka Springs, Ark.: Center on War and the Child, 1987.

Aldrich, G.H. & van Baarda, Th.A. (eds.), *Conference on the rights of children in armed conflict: final report of a conference held in Amsterdam, the Netherlands, on 20–21 June 1994.* The Hague: International Dialogues Foundation, 1994.

Amnesty International: Uganda *"Breaking God's Commands: the Destruction of Childhood by the Lord's Resistance Army",* AI Index: AFR 59/01/97, London, 1997.

Amnesty International: *"Old Enough to Kill But to Young to Vote", Draft Protocol to the Convention on the Rights of the Child on the Involvement of Children in Armed Conflicts,* AI Index: IOR/51/01/98, London, 1998.

Audoin-Rouzeau, Stéphane, *La guerre des enfants 1914–1918:* essai d'histoire culturelle. Paris: Armand Colin, 1993.

Boothby, Neil, *Reclaiming the killing field.* Durham, N.C.: Dukes University, 1990. (Dukes Perspective).

Boothby, Neil, Upton, Peter & Sultan, Abubacar, *Children of Mozambique: the cost of survival.* Washington, D.C.: U.S. Committee for Refugees, 1991. Issue paper.

Braeckman, Colette, Dossier: *Mozambique. In: Les Enfants du Monde.* Paris: Comité Francais Pour le Fonds des Nations Unies pour l'Enfance. 1993:116.

Brett, Rachel, *Involvement of children in armed conflicts: report on the third session of the working group to draft an optional protocol to the Convention on the Rights of the Child on involvement in armed conflicts 20–30 January 1997.* Geneva: Quaker United Nations Office, 1997.

Cairns, E: *"Children and Political Violence", Oxford:* Blackwells, 1996.

Campbell, Jim, *too young to serve: the consequences of a lost childhood.* Bristol: University of Bristol, 1997. (School of Policy Studies).

Cape Town annotated principles and best practice on the prevention of recruitment of children into the armed forces and demobilization and social reintegration of child soldiers in Africa. Cape Town: UNICEF. [report from a symposium in Cape Town 30 April 1997].

Child soldiers: a report from a seminar by Rädda Barnen and the Swedish Red Cross 12–13 February 1994, Stockholm, Sweden. Stockholm: Rädda Barnen, 1994.

Child soldiers: Development Education Forum. Lutheran World Federation. 1995:3.

Child soldiers: youth who participate in armed conflict. Washington, D.C.: Youth Advocate Program International, 1997. (Booklet no. 1).

Children, genocide and justice: Rwandan perspective on culpability and punishment for children convicted of crimes associated with genocide. Washington, D.C.: Save the Children Federation. USA, 1996.

Children in combat. In: *Human Rights Watch Children's Project.* 1996(vol. 8): no. 1(G).

Children in the North-East War: 1985–1995. Jaffna: The University Teachers for Human Rights (UTHR), 1995. (Briefing no. 2).

Children of Mozambique's killing fields: a summary report. Eureka Springs, Ark.: Center on War and the Child, 1989.

Children of war: a newsletter on child soldiers from Rädda Barnen. 1995: no. 1– . Stockholm: Rädda Barnen, 1995– .

Children of war: report from the conference on children and war, Stockholm 31 May 2 June 1991. Lund: Raoul Wallenberg Institute of Human Rights and Humanitarian Law, 1991.

Children's war: towards peace in Sierra Leone. New York: Womens Commission for Refugee Women and Children, 1997. (field report, March 26–April 16 1997).

David, Paulo, *Enfant sans enfance.* Paris: Hachette, 1995.

Dodge, Cole P. & Raundalen, Magne, *Reaching children in war:* Sudan, Uganda and Mozambique. Bergen: Sigma Forlag, 1991.

Dutli, Maria Teresa, Captured child combatants. In: *International Review of the Red Cross.* 1990: no. 278 (September–October), p. 421–434.

Dutli, M-T., and Bouvier, A., "Protection of Children in Armed Conflicts: the Rules of International Law and the Role of the International Committee of the Red Cross" *In: The International Journal of Children's Rights,* No. 4, 1996.

Frankel, Mark, Boy soldiers: a new and ruthless breed of warriors. In: *Newsweek.* 1995 (vol. CXXVI): no. 6, p. 3, 10–21.

Furley, Oliver, *Child soldiers and youths in African conflicts: international reactions.* Coventry: Coventry University, 1995. (African Studies Centre: occasional papers series,no 1).

Haapiseva, Jane, *Report of a field trip to Liberia, Mozambique and Zimbabwe on behalf of the Lutheran World Federation.* Geneva: Lutheran World Federation, 1994.

Hamilton, Carolyn & Abu El-Haj Tabatha, *Armed conflict: the protection of children under international law.* In: Netherlands: The International Journal of Children's Rights. 1997 (5): p. 1–46.

Human Rights: child soldiers. Brussels: Pax Christi International, 1993.

International Committe of the Red Cross: *"Position of the International Committee of the Red Cross on the Optional Protocol to the Convention on the Rights of the Child",* International Review of the Red Cross, Geneva, March 1998.

International Committe of the Red Cross: *"Plan of Action Concerning Children in Armed Conflict."* Stockholm: International Red Cross and Red Crescent Movement, Programme on Children Affected by Armed Conflict, 1997.

Invisible soldiers: *child combatants.* Washington, D.C.: The Defense Monitor. 1997 (XXVI): 4.

Koh, Aun, *Understanding volunteerism in child soldiers.* New York: Columbia University, 1995. Thesis.

Kosonen, Arto, *The special protection of children and child soldiers: a principle and its application.* Helsinki: University of Helsinki. The Institute of Public Law, 1987.

Lindsay-Poland, John, *Youth under fire: military conscription in El Salvador: a summary report.* Eureka Springs, Ark.: Center on War and the Child, 1989.

Los niños-soldados de Paraguay: investigacion sobre los soldados menores de edad. Asunción: Servicio Paz y Justicia (SERPAJ) Paraguay, 1996.

Lost boys: child soldiers in Southern Sudan. In: *Human Rights Watch/Africa.* 1994 (vol. 6): no. 11.

Louyot, Alain, *Gosses de guerre.* Paris: Robert Laffont, 1989.

Machel, Graca *"The UN Study on the Impact of Armed Conflict on Children",* UN Doc. A/51/306 of 26 August 1996 and Add. 1 of 9 September 1996, New York. (The countries covered include: Afghanistan, Bhutan, Burma/Myanmar, Burundi, Cambodia, Colombia, El Salvador, Ethiopia, Guatemala, Honduras, Israel/Occupied Territories, Lebanon, Liberia, Mozambique, Nicaragua, Northern Ireland, Paraguay, Peru, Philippines, Russian Federation (Chechnya), Rwanda, South Africa, Sri Lanka, Turkey, Uganda and Former Yugoslavia).

McCallin, Margaret, *The reintegration of young ex-combatants into civilian life: expert meeting on the design of guidelines for training and employment of ex-combatants, Harare, 11–14 July 1995.* Geneva: International Labour Office (ILO), 1995.

Muhumuza, Robby, *A case study on reintegration of demobilized child soldiers in Uganda.* Kampala: World Vision Uganda, 1995.

Muhumuza, Robby, *Girls under gun.* Kampala.: World Vision Uganda, 1997.

Muhumuza, Robby, *Gulu: the children of war: a report from a UK-ODA assisted programme in Uganda.* Milton Keynes, 1996.

Muhumuza, Robby, *The gun children of Gulu: the reluctant child soldiers in Joseph Kony's Lords Resistance Army (LRA) in Northern Uganda.* Kampala: World Vision Uganda, 1995.

Parker, Richard, *Iran's child martyrs: a summary report.* Eureka Springs, Ark.: Center on War and the Child, 1987.

Pint, Nathalie, *Child Soldiers. Colchester:* University of Essex, 1993. Dissertation.

International Save the Children Alliance, *"Promoting psychosocial well-being among children affected by armed conflict and displacement: principles and approaches",* Geneva, 1996.

Proxy targets: *civilians in the war in Burundi* – New York: Human Rights Watch, 1998

Raundalen, Magne & Dyregrove, Atle, *Reaching children through the teachers: a manual for helpers of war-affected children.* Bergen: University of Bergen, 1992.

Recruitment and use of children in the Gulf war: a summary report. Eureka Springs, Ark.: Center on War and the Child, 1988.

Refugee children: guidelines on protection and care. Geneva: UNHCR, 1994.

Reis, Chen, *Trying the future, avenging the past: the implications of prosecuting children for participation in internal armed conflict.* New York: Columbia Human Rights Law Review, 1997. (Vol.28, no 3, Spring 1997).

Ressler, Everett, Tortorici, Joanne Marie & Marcelino, Alex, *Children in war: a guide to the provision of services.* New York: UNICEF, 1993.

Rights of the child: report of the working group on a draft optional protocol to the Convention on the Rights of the Child on involvement ofchildren in armed conflits on its second session. Geneva: UN Economic and Social Council (ECOSOC). (E/CN.4/1996/102, 21 Mar. 1996).

Rompan filas: *fotografias de Jorge Sáenz.* Asuncion: Servicio Paz y Justicia (SERPAJ) Paraguay, 1996.

Runyan, Curtis, Throwing children into battle. Washington: Washington Post, 1997. (May/June issue of *World Watch excerpt* in May 8, 1997, p.26).

State of the world's children 1996. New York: Oxford University Press for UNICEF, 1995.

The Scars of death: children abducted by the Lord's Resistance Army in Uganda. New York: *Human Rights Watch,* 1997.

Stop using Child Soldiers! London: Coalition to Stop the Use of Child Soldiers and International Save the Children Alliance, 1998.

Tolfree, David, *Restoring playfulness: different approaches to assisting children who are psychologically affected by war or displacement.* Stockholm: Rädda Barnen, 1996.

Uganda: land of the child soldier: a summary report. Eureka Springs, Ark.: Center on War and the Child, 1987.

Victims of violence: minors and the armed conflict in Colombia, system of monitoring and vigilance, childhood and its rights. Bogota: Public Ministry Ombudsman's Office, 1996.

Woods, Dorothea (ed.), *Child soldiers: the recruitment of children into armed forces and their participation in hostilities.* London: Quaker Peace and Service, 1993.

Woods, Dorothea, *Children bearing arms: a form of child labor?: a summary report.* Eureka Springs, Ark.: Center on War and the Child, 1990.

Woods, Dorothea, *Children bearing military arms in Latin America: a summary report.* Eureka Springs, Ark.: Center on War and the Child, 1990.

Woods, Dorothea, *Children bearing military arms in Uganda:* a summary report. Eureka Springs, Ark.: Center on War and the Child, 1990.

Woods, Dorothea, Lebanon: *children in armed conflict from 1975–1989: a summary report.* Eureka Springs, Ark.: Center on War and the Child, 1990.

Index

For countries, see also Annex 1 (page 205)

abduction
    see kidnapping

adventure: 57, 60

advocacy: 10, 113, 116–118, 122–123, 129, 149, 161, 174, 189, 195, 274–275

Afghanistan: 12, 16, 24, 35–37, 48–49, 52, 61, 64–66, 68–69, 71, 73, 75, 78–79, 81–82, 87, 90–91, 93, 95, 98, 103, 105–106, 113–114, 116–117, 133–134, 136, 141, 210, 226

age:
    of criminal responsibility: 97, 108–109, 188, 203–204
    distribution: 32
    of majority: 14, 30, 117, 163, 167–168, 202
    see also minimum age

Albania: 24, 210, 226

alcohol: 88, 91, 93, 98, 145, 194, 260

Algeria: 24, 210

Angola: 24, 56, 74, 210

Argentina: 163, 210

atrocities: 25, 79, 98, 169, 172, 203, 261

Australia: 91, 157, 210

Austria: 163, 210

Azerbaijan: 24, 210

Bangladesh: 24, 210

Bhutan: 12, 16, 30, 47, 49, 52, 71, 93, 98, 100, 210

birth records: 42–43, 174, 186, 199

bribes: 47, 72, 80

article 37: 97, 107, 167, 202
article 38: 14, 25, 31, **158–161,** 163, 167, 202, 273
article 39: 151
article 40: 107
article 41: 161
optional protocol: 164–166, 168, 183, 203, 274, 277–278

crime: 21, 25, 26, 68, 106, 115, 136, 140

criminal activities: 131, 172, 263
    behaviour: 140
    proceedings: 108
    see also age of criminal responsibility

culture: 57–61, 64, 79, 81, 111, 116–118, 122, 124, 137–140, 146–147, 181, 190, 196, 262
    see also tradition

death penalty: 107, 109, 167, 169

delinquency:
    see crime; criminal activities

dehumanisation
    see brutalisation

denial (of children being armed): 19, 21, 27, 115, 119–120, 124

desertion: 32, 71, 101, 114, 116, 128, 154, 155, 194
    see also escape

detention: 61–62, 106–107, 169, 184, 195
    arbitrary 62, 173, 187, 200
    see also prisoners

disability: 88, 114–116, 145, 191, 194, 263

disadvantaged: 52, **69–74,** 131, 181

drugs: 93, 98, 131, 142, 145, 194, 260

El Salvador: 12, 16, 19–20, 30, 33, 36, 43, 46, 48–52, 55–57, 60–63, 65–67, 71–72, 75, 78–79, 82–83, 87–88, 91, 93–96, 98, 100, 103, 106, 108, 116–117, 119, 125, 134, 136, 139, 176–180, 214

employment: 36, 54, 66, 82, 96, 105, 112, 121, **133–136,** 148, 154, 167–168, 196–197, 202, 264

escape: 48, 51–52, 56, 68, 71, 90, 92, 100, 155, 194
    see also desertion

ethics: 41, 150, 189

Ethiopia: 12, 16, 19, 24, 35, 43, 47–50, 52–54, 56–57, 60–61, 64, 66, 68, 71, 73, 78–79, 82–83, 87, 90, 93, 95–96, 98, 100–101, 103, 106, 120–121, 132, 153, 156, 214

ethnicity: 42, 49, 61–62, 64, 70, 74, 118, 125, 137–140, 260, 263
    see also indigenous, minorities

ethnic groups: 61, 176

execution: 100

family background: 79–80, 180
    reunification: 108, 121, 126–133, 180, 190, 196
    substitute: 119, 132

forced recruitment
    see recruitment

Former Yugoslavia: 12, 19, 30, 49, 60, 64, 73, 79, 87, 119, 179

gender: **81–85,** 123, 137, 188, 259–261

girls: 28, 42, 64, 68, **81–85,** 92–93, 145–146, 191, 194, 197, 220, 259–260, 263, 265

Germany: 163, 214

Guatemala: 12, 16, 34, 43, 48–49, 52, 54, 57, 65, 67, 71–72, 84, 87–88, 93, 95, 97–98, 101, 103–104, 106, 108, 115, 139, 173, 176, 178, 180, 214

handicap
    see disability

harassment: **61–63,** 78, 169, 173, 184, 187, 194–195, 202

health: 101–105, 107, 131, 144, 151, 167, 188, 194, 202, 263

history: 13, 20, 56

Honduras: 12, 16, 30, 41, 48, 64, 71, 85, 88–90, 92–93, 98, 100, 116, 146, 176, 180, 216

ideology: 9, 57, 64–65, 79, 91, 117, 119, 144, 262

imprisonment
    see prisoners

indigenous: 49, 70, 138, 260
    see also ethnicity; minorities

indoctrination: 65, 91, 121

Indonesia: 24, 216

injuries: 10, 63, 75, **101–105,** 114, 145, 194, 261

injustice: 25, 125, 149

institutionalisation: 70, 131, 190, 197

International Catholic Child Bureau: 11, 123, 135, 147

International Committee of the Red Cross, ICRC: 21, 121, 187, 202, 275
  see also Red Cross

international humanitarian and human rights law: 21, 25, 97, **158–169,** 175, 188, 204,
  280

Intifada: 12, 16, 30, 60–64, 73, 79, 99, 106, 108–109, 137, 139, 144

invalidity
  see disability

Iran: 24, 216

Iraq: 24, 216

Israel/Occupied Territories
  see Intifada

juvenile justice: 107, 115, 140, 191

kidnapping: 26, 50, 55, 79, 99, 123

Kosovo: 258

landmines
  see mines

Lebanon: 12, 16, 24, 32, 48–49, 61, 64, 66–67, 69, 73–75, 79, 81, 85, 90, 97, 103–104,
  106, 108, 117–118, 125, 130, 145, 154, 173, 218

length of service: 123

Liberia: 12, 16, 20, 24, 32, 36, 48–49, 51, 57, 60–62, 66–67, 73, 75, 78–80, 92–93,
  95–96, 98, 100, 112–113, 121, 144–145, 154, 173, 179–180, 218

Lutheran World Federation: 11

Machel Study
  see Study on the Impact of Armed Conflict on Children, Machel Study

martyrdom: 64
  see also suicide

media: 20, 71, 118, 273

Mexico: 24, 218

militarisation: 64, 127, 174–175, 201, 265, 273

military
  court: 108
  hospitals: 103
  law: 100, 115, 188, 203–204
  prisons: 106
  schools: 166, 188, 204
  service: 42, 46–47, 52–53, 69–70, 73, 82, 96, 116, 138–139, 155, 174, 185, 201
  training: 26, 56, 80, 84, 93–94, 100

militias: 14, 16–17, 19, 32, 34, 36, 47, 53–54, 161, 173, 185, 199–200, 265

mines: 95, 97, 102–104
  clearance: 104
  injuries: 102, 104

minimum age: 14, 31, 42, 157, **161–169,** 172, 175, 183–184, 186, 198, 200, 202–203, 259, 277

minorities: 42, 118, 125, 138, 188, 201, 203
  see also ethnicity; indigenous

monitoring: 27, 114, 178, 184, 187, 200, 265

Mozambique: 12, 16, 30–32, 36, 42–43, 49–50, 52, 54, 56, 71, 73, 79–80, 85, 92–95, 98, 100, 103–104, 106, 108, 120–121, 123, 130, 134, 139, 218

Myanmar
  see Burma

national law: 14–15, 25, 31, 43, 157, 161, 178, 181, 186, 200, 202, 259

Netherlands: 91, 157, 163, 218

New Zealand: 218

Nicaragua: 12, 16, 30, 34, 49, 57, 66–67, 71, 73–74, 87, 95, 98–99, 108, 121, 137, 141, 218

non-governmental organisations, NGOs: 12, 27, 121, 147, 172, **174–176,** 178, 180–181, 184, 186–188, 190–191, 195, 198–202, 275, 277–278

Northern Ireland: 12, 33, 95, 99, 108, 156

Oman: 220

Optional protocol
see Convention on the Rights of the Child, CRC

orphans: 56, 66, 128, 180

Pakistan: 24, 220

Papua New Guinea: 24, 220

Paraguay: 12, 16, 20, 30, 43, 48–49, 71, 88–89, 99–100, 154, 176, 220

payment: 48, 66–67, 99, 115, 265

peace
agreement: 265
education: 34, 137

peer pressure: 60, 80

penalisation
see punishment

Peru: 12, 17, 24, 32, 36, 49, 51, 54, 56–57, 70–71, 80–82, 85, 90–93, 95–96, 99, 108, 115, 153–154, 176, 220

Philippines: 12, 17, 24, 63–64, 66–67, 73, 79, 84, 91, 93, 95, 100, 103, 106–107, 113–114, 116–117, 125, 144, 156, 180, 220

press-ganging: 50–52, 71, 198
see also forced recruitment

prevention: 25, 62, 112, 117, 124, 136, 160, 162, 164, **171–181,** 186–191, 198–203, 262, 274

prisoners: 80, 156
of war: 106–107, 261
see also detention

prosecution
see juvenile justice

prostitution: 129, 146

publicity
see media

punishment: 25, **99–101,** 140, 188, 194, 203–204

Quakers: 11, 91, 275, 278

quota enlistment: 42, 46

Sri Lanka: 12, 17, 24, 56, 60–61, 63–64, 66–69, 73, 75, 84–85, 93–95, 98, 103, 114, 128, 154, 222

street children: 56, 79, 201

Study on the Impact of Armed Conflict on Children, Machel Study: 10–12, 14, 25, 126, 164

substance abuse
see drugs

Sudan: 10, 24, 43, 48–49, 51, 54, 56, 65, 73–74, 79, 178, 222

suicide: 64, 85, 88–89, 97, 152, 154, 194–195
see also martyrdom

Switzerland: 163, 222

Tajikistan: 24, 222

torture: 61–63, 92, 106, 144, 187, 200

tradition: 49, 60–62, 66, 138

trauma: 92, 105, 134, 147–151, 197

Turkey: 12, 17, 24, 42–43, 46, 49, 54, 56, 62, 65, 67, 79, 81, 84, 87, 91, 93–94, 98, 103, 105–106, 108, 113, 116, 222

Uganda: 12, 17, 24, 32, 49, 51, 62, 78–79, 83–85, 90, 92–95, 98, 100–101, 103, 108, 132, 136, 141, 146, 223

unaccompanied children: 56, **78–80,** 121, 128, 133, 139–140, 147, 180, 194, 201, 260, 263

unemployment
see employment

United Kingdom: 12, 24, 224
see also Northern Ireland

United Nations High Commissioner for Refugees, UNCHR: 175, 178, 186–187, 190, 198–202, 275

United Nations Children's Fund, UNICEF: 34, 113, 121, 131, 147, 175, 178, 186, 190, 198–202, 295

USA: 129, 140, 158, 166, 208, 224

Uruguay: 163, 224

Vietnam: 105, 224, 275